EAA
"HOW TO"
SERIES

EAA® AIRCRAFT BUILDING TECHNIQUES
WOOD

AF835598

Contents

Compiled and prepared
by Paul Poberezny
and S.H. "Wes" Schmid

Published July 1994. Reprinted 1998.

Moisture Content
Wood's Vital Factor

By Bob Whittier, EAA 1235

MANY SPECIES of wood can be and indeed have been used in the construction of small airplanes. But, no matter what kind is chosen and how carefully it is selected, maximum strength and durability will not be attained unless the material is handled with a thorough working knowledge of the principles of moisture content. For moisture content is the vital factor in the art of wooden aircraft construction.

A living tree contains an enormous quantity of moisture, sometimes as much as 115 percent. Moisture content is the weight of the water contained in the wood expressed as percentage of the weight of completely dried wood. One illustration of how much liquid there is in a tree is given in the statement that if an average sized house were to be made of very green wood, it would contain about 500 gallons of water!

Trees have been found which had as much as 300 percent moisture content, while at the same time the heartwood of some coniferous trees is known to have as little as 30 percent moisture content. Moisture content is not uniform throughout a tree, it being a common occurrence to find wood in the lower part of the trunk to be wetter than that higher up. It is commonly thought that trees contain more moisture in the warm months when the sap is "up", but this is a complete misconception. Tests show that trees cut in the winter contain as much water as those felled in the summer.

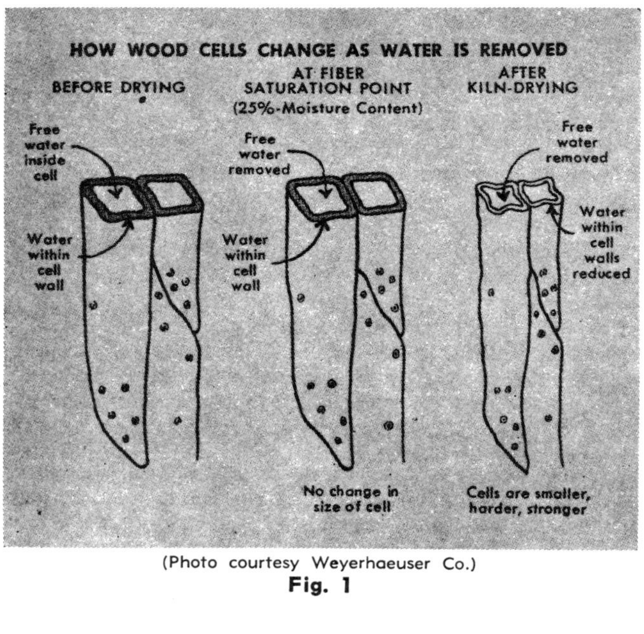

(Photo courtesy Weyerhaeuser Co.)
Fig. 1

The moisture in wood takes two forms: 1. Free water within the open spaces in the cells. 2. Moisture in the walls of the cells. In Fig. 1. is shown how wood cells change as water is removed. First, the free water inside the cell goes out, which reduces the wood's moisture content to approximately 25 percent. The cell structure does not undergo any change. Second, water leaves the cell walls, and as this occurs the cells contract, harden and stiffen. The wood then definitely gains strength and shrinks noticeably. When wood has been dried to the point where moisture has gone out of the cell walls, it develops what might be called a "closed pore" structure. When common building lumber is kiln-dried to an average moisture content of 19 percent, for example, its bending strength increases by 22 percent, the strength in horizontal shear becomes 10 percent greater, stiffness increases by 9 percent, and crushing strength goes up a thumping 32 percent. In addition, **once the cell walls lose their moisture** and enter the "closed pore" state, they develop appreciable resistance toward re-absorbing moisture. Seasoned lumber is therefore more stable dimensionally, aiding accurate working. Fig. 2 shows in graphic form the influence of moisture content on strength.

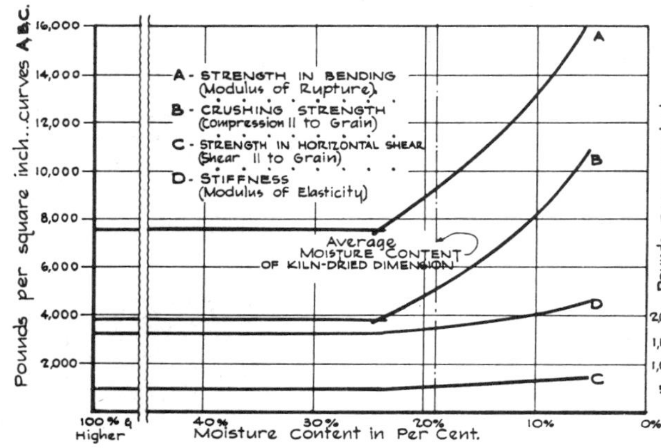

Graph based on Douglas Fir using Forest Products Laboratory (U.S. Department of Agriculture) literature.
Fig. 2 Strength increases as wood dries out.

While lumber can indeed be air-dried, the process has shortcomings for many applications, such as general building and aircraft construction. It is slow, and due to the normal variation of moisture content in pieces of green wood, there is apt to be noticeable variation in the moisture content of wood seasoned by air-drying. Kiln-drying is faster and more uniform, but the drying rate must be skillfully controlled as too-rapid drying can result in splitting, cell collapse, and a condition of non-uniform drying called case-hardening, in which the inside is wet and soft and the outside hard and dry, leading to warpage when the wood is sawn.

The process of cutting and kiln-drying aircraft grade wood is scientifically controlled to guarantee safe, uniform material. Green, wet wood when exposed to favorable air conditions is apt to develop surface molds and stains rather quickly, so it is common to dip freshly-cut wood into a fungicidal bath after the log has been sawn. This protects the wood until it can be dried. Wood dries out at its ends more quickly than in the core of the stock, and the rapidity of kiln-drying as compared to air-drying aggravates the tendency for end checks to form. Ends of planks are therefore covered with sealant coatings formulated both to retard end drying and to resist kiln heat. The process of selecting, cutting and treating aircraft wood is covered in great detail in the government publication, ANC-19, Wood Aircraft Inspection and Fabrication.

Commercial drying kilns are often rather large buildings, such as the one in Fig. 3, specially built for the purpose and incorporating features to control temperature, humidity, air circulation and ventilation to best advantage. Wood always shrinks more across the grain than it does along the grain. Since drying time is a function of surface area exposed to the air, it would take far more time and fuel to kiln-dry solid logs than sawn boards. For this reason, wood is sawn to the desired rough dimensions before being put in the kiln. Due to the effect of grain direction and board cross sections on the shrinking pattern, wood has to be sawn out of logs in such a way as to obtain aircraft stock with minimum deformation tendencies. By far the least deformation, straightest grain and most strength is obtained from the cut occupying the 9 o'clock position in Fig. 4. Obviously, rather few pieces can be obtained from a log by this cut as compared to some of the others. This is yet another reason why selected aircraft grade wood is expensive . . . but since lives are at stake, the care is warranted. Trimming to final dimensions is always done after the wood is dry.

(Photo courtesy of Forest Products Laboratory)
Fig. 3 Drying Kiln.

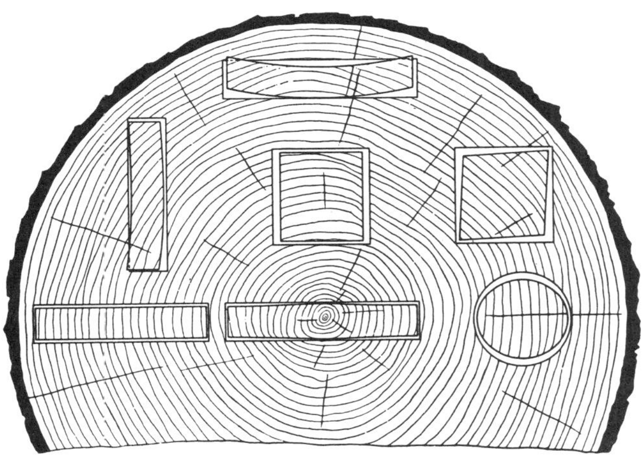

Fig. 4 Effects of radial and tangential shrinkage on the shape of various sections in drying from the green condition.

Aircraft lumber normally has a moisture content of 12 percent. It should not fall below 8 percent, as totally dry wood can be brittle and may absorb so much water from glue as to produce weak joints. Also, as total dryness is practically never found in service conditions, totally dry wood would inevitably expand, straining the glue joints seriously. For propellers, where moisture content directly affects the balance, it has been common to use dryer wood, in the 5 to 7 percent range, but this is the only exception. When water-mix glues such as casein are used, a stiffer mix having perhaps 10 percent less water may be used, because the dry wood will draw less water from a thick glue mix than from a thin, watery mix, with less chance of weak or starved joints. When moisture content is over 12 percent there are problems of glue adhesion, weight, dimensional changes and glue-line fractures when the wood becomes dryer in arid climates, and adhesion of finishes.

In factories building wooden airplanes extensive steps are taken to control wood's moisture content. Material in storage may be kept in buildings or rooms having a controlled atmosphere, and work rooms may be equipped with either humidifying or dehumidifying provisions. Lumber stored in the arid Southwest has been found to have a moisture content of as low as 6 percent during summer months, while in damp winters lumber in the Pacific Northwest has had as much as 26 percent moisture content.

The amateur builder in a steam-heated winter workshop may not realize how dry his wood can become; for about $10 one can get a good hygrometer at the clocks-and-instruments counter of a department store, and this will give a check on the shop's atmosphere. If it is dryer than about 12 to 15 percent, wood kept there for periods of more than about a week can become too dry. The simplest of remedies is to moisten the air by means of a common radiator-type humidifying pan, or a towel hanging from a bucket of water, or even, in stubborn cases of dry air, a jet of steam from a radiator or a very fine jet of water from a pinhole, directed at a flat plate so that it atomizes upon contact. As a rule, moisture content control is a minor problem in temperate United States, although sometimes the dampness becomes a problem in Gulf Coast states. In that locality, kiln-dried aircraft wood shipped in from a "dry" state might well be kept in the least-humid part of the house. The pilot light of appliances such as gas hot water heaters are known to take a certain amount of water out of a basement's air, which suggests the possibility of some kind of flame arranged so as not to heat the room and to create no fire hazard. And, of course, home style dehumidifiers are on the market. The best way to be sure in a notably wet or dry climate is to keep tabs on the atmospheric humidity with a hygrometer, and if the wood is exposed for long to such air something can be done along these lines.

In production work, electric moisture-content meters such as in Fig. 5 are used. The probes are held against the wood and as wood's electric conductivity is a function of the moisture content, a direct reading is obtained. Some knowledge is needed to use such a meter effectively; for example a piece of wood taken out of a kiln too soon would appear to have an acceptably low surface moisture content, in spite of the fact that it might still have a lot of internal moisture. Some time later when this moisture had had time to work to the surface, a much higher moisture content reading would be obtained. Or, a very short piece of wood, which will dry out faster than a long piece, might be tested and on the basis of its low moisture content reading, a still-green batch of long boards could be released for use by an ignorant operator. When you buy certified aircraft grade stock, you know that tricks like this have been taken into account by expert lumbermen. Electric meters are too specialized to have any place in the amateur's inexpensive workshop equipment, but you get the benefits of one when you order aircraft grade spruce.

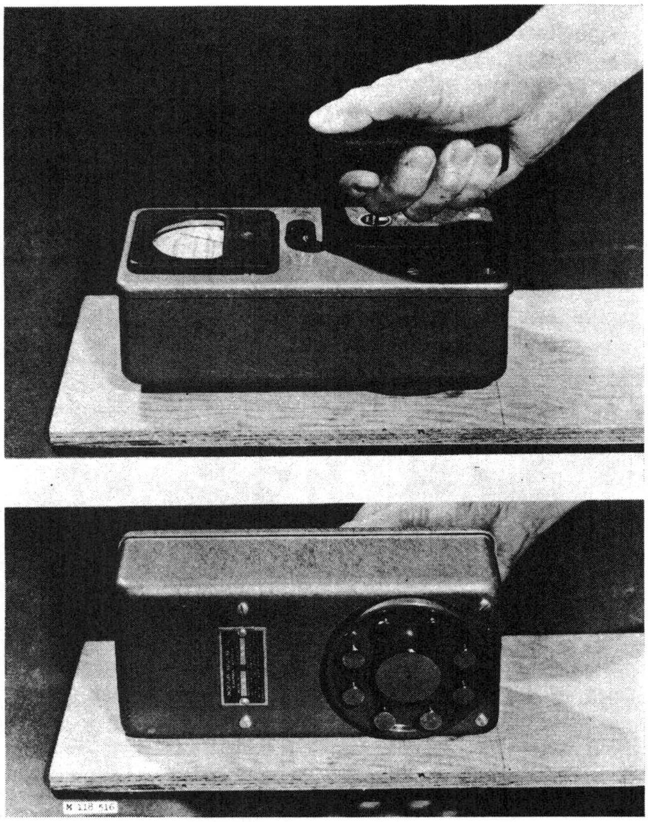

Fig. 5 Electric Moisture Content Meter.

It is possible to get fairly good moisture content measurements with the simple oven-dry method. Select a representative sample of the wood, preferably from two or more feet in from the stock's end, as the ends take up and give off humidity more rapidly. A short test piece is best, as it will dry out from its ends faster. A cube one inch square would serve nicely. Cut it smoothly and accurately, and remove splinters. Weigh it immediately, as so small a piece will start to dry at once. Small postal scales are not sensitive enough so use a light chemist's scale or equivalent. An accuracy of one half of one percent is desirable.

Put the sample in an oven between 212 and 221 deg. F. and dry until a constant weight is achieved. When no more weight change is discernable between weighings the wood is totally dry. Obtain its oven-dry weight. Divide the loss-in-weight figure by the oven-dry figure and multiply the result by 100 to get the percentage of moisture in the original sample thus:

$$\%\ \text{moisture} = \frac{(W - D)}{D} \times 100$$

Where W = original weight
D = oven-dry weight

A typical example follows. Assume an air-dry sample of Sitka spruce 2" x 2" x 1", weighing 30.8 grams. After drying it weighs 27.5 grams. Find the moisture content:

$$\%\ \text{moisture} = \frac{(30.8 - 27.5)}{27.5} \times 100 = \frac{330}{27.5} = 12\%$$

It is important to understand the characteristics of various types of coatings as regards their moisture-excluding properties. Always remember that any protective coating will keep moisture **in** wood as well as it will keep moisture **out**; if wood gets wet through a structural fault or detail which admits rainwater or condensed water, the coating on the surrounding wood will tend to hold it in the wood and thereby create a very pronounced rot hazard.

What we want of any protective coating is an ability to keep the wood surfaces from absorbing water and humidity after we have taken the care to eliminate cracks and crevices which create traps. In general if a coating over properly-dried wood serves to appreciably slow down the rate of absorption of atmospheric moisture, the wood will not become wet enough to swell and crack glue lines, or warp the structure, or be susceptible to rot. Rot can not and will not take place when wood's moisture content is kept below about 20 percent.

It's easy to become confused as to the properties and purposes of various wood coatings on the market today. No wood finish yet developed is entirely impermeable to moisture; no matter what kind is used, if the exposure period is long enough, the wood will eventually absorb moisture. We do not often have wet weather that lasts long enough for wood to become really wet under the coatings we do have. In tests, samples exposed to 100 percent relative humidity took from 10 to 90 days to reach a stage of moisture content equilibrium, these figures being for the poorest and best coatings respectively. In the tropics, wooden planes used to give more trouble from fungi attacking the vegetable and animal glues than the wood itself. With the advent of glues that repel fungi, the life of planes in the tropics increased greatly. Today, if the wood is given a fungicidal treatment prior to final finishing and the structure is designed to provide free and general circulation of air, and perfect drainage of any free water, long life can be expected.

Here is how the finishes situation works out:

1. Deeply penetrating finishes, such as water-repellent preservatives. These afford rather low moisture-excluding effectiveness but may have useful repellency to shed splashed water. Their primary purpose is to act as carriers for fungicidal chemicals. The more deeply they penetrate, the more thorough is the fungicidal protection. To reach all possible areas, application to finished structures is usually by dipping for not less than three minutes. For aircraft use, a penetrating preservative conforming to government specification AN-P-76 is suggested. Some firms in the penetrating preservative field are Protection Products Manufacturing Co., Kalamazoo, Mich., Wood Treating Chemicals Co., St. Louis, Mo., and Chapman Chemical Co., Memphis, Tenn.

2. Slightly penetrating finishes that form little or no coating over the surface of the wood, such as so-called "sealers". Their primary purpose is to fill surface pores and render it non-absorptive to finish coatings. Water repellency is low for single coats, but two or more can provide moderate protection. In the everyday market, the plywood sealers sold in lumber yards would be representative of this type. In the aircraft industry, Army Sealer 3-186, types II and III, are suggested.

3. Porous coatings, such as dull camouflage paints and "sanding surfacers" used to build up wood surfaces prior

to high-gloss paint jobs. Low moisture-excluding effectiveness.

4. Varnishes. When applied with appreciable film thickness, usually two, three or more coats such as to kill all fuzz and give a glossy surface, will give reasonably high moisture excluding effectiveness. No good when exposed to sunlight as that makes varnish dry out and crack in a short time, but acceptable for sheltered internal areas.

5. Pigmented coatings of a non-porous nature such as semi-gloss and gloss enamels. Higher moisture exclusion than similar coatings without pigments. The trouble with all pigmented finishes for aircraft woodwork is that it renders it difficult to spot incipient rot and cracks in wooden structural members, due to the opacity. Rot can get a dangerous foothold before the covering paint will peel or crumble enough to give warning of it.

6. Varnishes with aluminum pigmentation. Very high moisture excluding effectiveness even in light, thin coats. Best results had with this preparation when the first coat consists of varnish with aluminum powder and is applied as a primer, with subsequent second and perhaps third coat of clear varnish to seal the surface. Applying plain varnish for the first few coats, followed by an aluminized finish coat, results in an appreciable drop in sealing power. This coating is highly recommended for spar butt ends, inside propeller hub holes and other places where end-grain is exposed to moisture. It is also used in landing gear wheel wells and other places very prone to receive splashing, but is not often used as an overall coating because it prevents visual inspection of the wood.

The ideal treatment for the structural members of an all-wood airplane therefore would be to apply fungicidal penetrating preservative as insurance against the development of rot if water does get into the wood, followed by a gloss varnish job to seal the wood as much as possible against entry of moisture. But no kind of finish will preserve an airplane's structural integrity unless the wood was originally kiln-dried to a suitable low moisture content level, and the plane was assembled in such a way as to avoid all chance of trapping water in the structure.

Aside from safety, sealing the wood has an important purpose in regard to performance. Years ago, some all-wood airplanes built by pioneer airmen absorbed so much moisture during wet weather that they would not fly! In a low-powered all-wood sportplane, it is just as desirable to minimize moisture absorption.

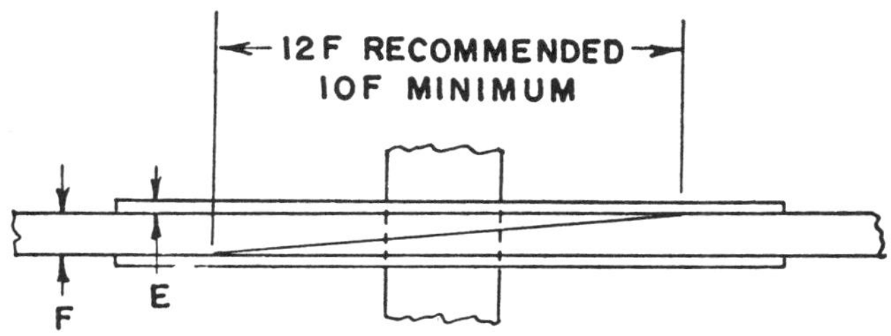

AT A SPAR

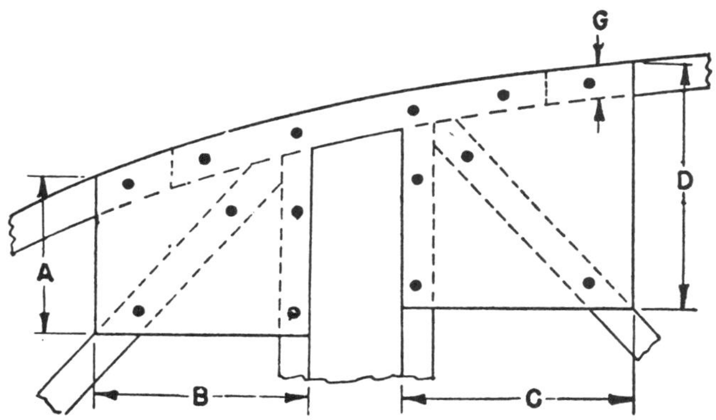

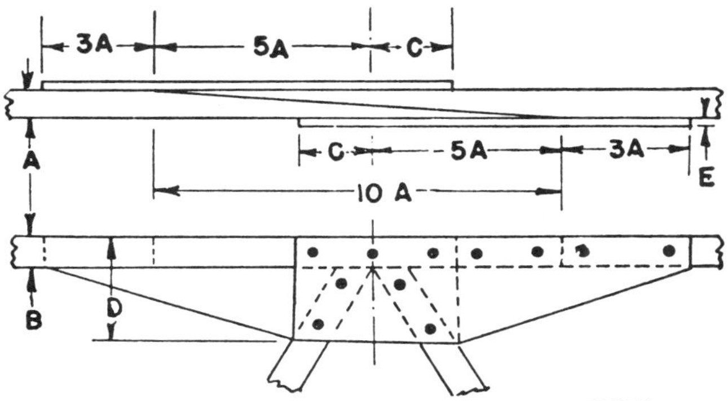

AT A JOINT TO TRUSS MEMBERS

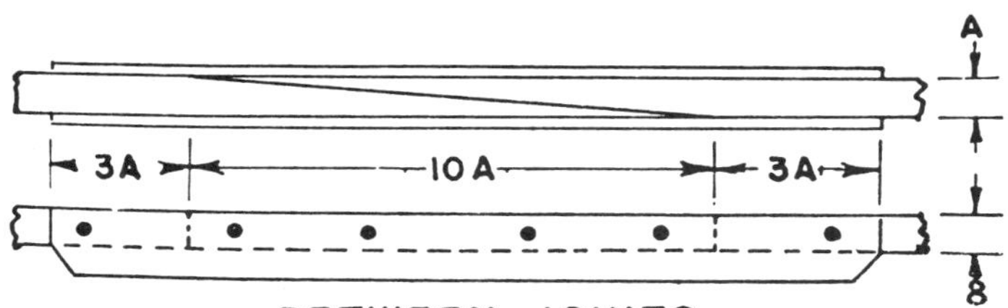

BETWEEN JOINTS

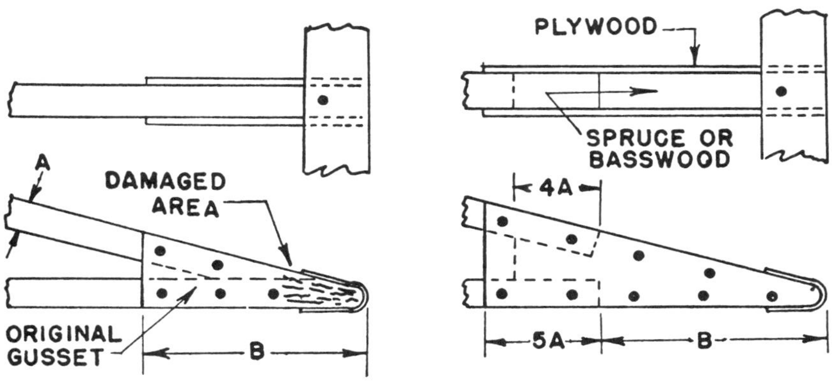

AT TRAILING EDGE

A, B, C, D, E, F, G = ORIGINAL DIMENSIONS
REINFORCEMENT PLATES SHALL BE PLYWOOD GLUED AND NAILED
THE DIRECTION OF THE FACE GRAIN MUST BE THE SAME AS ORIGINAL GUSSET.

Love That Picca Sitchensis

(Reprinted from USDA Forest Service Brochure)

FEAR NOT, fellow homebuilder of aircraft. Picea sitchensis is not a rare tropical disease; it is the botanical term for that wonderful aircraft wood, Sitka spruce. Sometimes called "the glorious Sitka spruce," because it is a tree of splendid form, this species produces wood of high commercial value. Larger than any other spruce in North America or perhaps in the world, its size is comparable to that of the huge Douglas fir. The great size of these trees makes it possible to obtain clear lumber of almost any desired width and length of grain.

Up to the first World War, Sitka spruce was used largely for local consumption. During that time however, its exceptional fitness for aircraft construction was discovered and it became prominent both in this country and abroad.

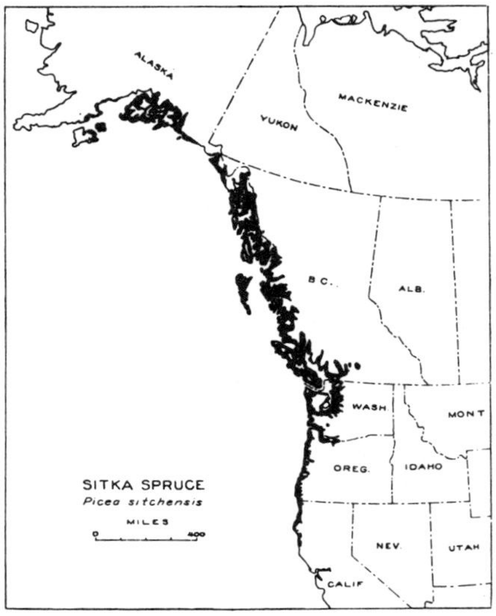

Sitka spruce is rarely found more than 40 miles from the coast.

Sitka spruce grows in a strip along the Pacific coast from northern California to a little beyond Cook Inlet in Alaska, where in the coastal forests this species with western hemlock makes up the larger part of the stand. The latest available figures for Sitka spruce show the amount to be roughly as follows: Alaska, 17.7 billion board feet; British Columbia, 6.2 billion; Washington and Oregon, about 8 billion; with 0.2 billion in California.

Although Sitka spruce occasionally forms pure stands, it generally grows in mixture with other species. In Oregon and Washington it is associated with Douglas fir, grand fir, western hemlock, and western red cedar. Forest grown, trees are tall and straight, with open conical crowns. Those approaching maturity have long cylindrical trunks, frequently free of branches for 40 to 80 feet, and often with spreading bases. Mature trees in Washington and Oregon commonly measure from four to six ft. in diameter and more than 200 ft. in height and attain an age of 400 to 450 years. In Alaska, average dimensions of mature trees are somewhat less.

The wood of Sitka spruce is straight grained and comparatively uniform in texture, moderately light in weight, moderately soft, moderately stiff, moderately low in resistance to shock, and easily worked. Very strong for its weight, it possesses greater strength and toughness than any wood of similar weight. It is not difficult to kiln dry and is moderately resistant to decay. The wood has no distinct odor and contains very few resin ducts.

The principal uses for Sitka spruce are lumber and paper pulp. Material sawed for lumber is used for construction purposes, but most of it goes into remanufactured products, including boxes and crates, planing mill products, sash, doors, blinds, and general millwork, refrigerators and kitchen cabinets, woodenware, novelties, furniture, fixtures, boat building and, of course, aircraft building. A small amount is used for sounding boards for pianos and other musical instruments.

Because it has unusual strength for its weight and can be obtained in clear straight grained pieces of large size and uniform texture with but few hidden defects, Sitka spruce is by far the most important wood for aircraft construction. It has gone into various parts of the planes of many famous aviators, and is the basis for almost all wood aircraft, both commercial and homebuilt. The requirements for aircraft construction are so exacting, however, that only about five percent of the cut of Sitka spruce can be utilized for this purpose.

On the Use of Douglas Fir as a Substitute for Spruce

By Stan Hall, EAA 10883*

HAVE YOU looked into the going price for spruce lately? Scandalous. And have you tried to find it in your local lumber yard? Impossible. You have to send away for it. And sometimes you don't get what you asked for.

Why not forget about spruce and use vertical grain Douglas fir instead? A lot of us do. Fir costs only about one-fifth as much as spruce, is beautifully straight grained, can be procured in unusually long lengths and best of all, can be purchased at most sizable lumber yards right there in your own town.

Why isn't fir more widely used among the experimental aircraft builders? For two reasons, I believe. One is that spruce is the traditional wood for aircraft and tradition, being what it is, is difficult to break with. People think spruce, so spruce it is.

Another reason is that few people take the trouble to compare other woods with spruce on an engineering basis and thus information about them is not as widespread as it ought to be.

Before developing a case for fir, let's look a bit closer at spruce. Why does it cost so much? For one thing, the same small stand of spruce growing in a narrow band along the northern Pacific coast has been supplying the needs of U.S. aircraft builders since the days of the Wright Brothers. In addition, spruce finds considerable use in the construction of pleasure boats. As anybody knows, any time you have a limited supply and a heavy demand, you have high prices. This is one reason spruce is uncommonly expensive in comparison to other suitable woods.

A third reason is that we aircraft builders insist on the best spruce we can find. It has to be stamped "certified aircraft" or we turn up our noses. This stamp is put on there by an inspector who selects the best pieces from the lot — and adds his pay to your bill.

By the time the spruce is inspected and stamped, shipped to a jobber (the man **you** deal with), wrapped or crated and shipped across the country to your home via train and truck, you have a sizeable bill on your hands.

*Professional Aeronautical Engineer; Designer of sailplanes "Cherokee II", "IBEX", et al; Chairman, Research Committee, Soaring Society of America, Inc.

However, as one advertiser for spruce put it, you "deserve only the best." You also deserve a new Cadillac, but I doubt if you have one.

Your new aircraft may deserve the best, but it likely doesn't need it. Any professional airplane designer who uses the highest quality material available, regardless of cost or real need, soon finds himself out of a job. And I think there is a lesson here.

If your aircraft really **needs** spruce and you are willing to go to the trouble of finding it and paying for it, by all means use it. But do you really need it? Before you decide, let's look at some trade-offs between spruce and vertical grain Douglas fir. It may be an eye-opener. We've already talked about availability and touched upon cost, so let's look now at the physical characteristics of the two woods. This is summarized in the following table. You won't need the whole table to help you make up your mind but I've put it all there for the sake of completeness and because the data are readily available; it comes right out of ANC-18, "Design of Wood Aircraft Structures."

Let me emphasize that we are talking here about **Douglas fir,** not other kinds of fir such as California red fir, Noble fir, Pacific silver fir and white fir. Don't make a mistake and get the wrong kind. And be sure the Douglas fir is **vertical grain**. If you don't know what this is, your lumberman will.

Let's look at selected elements in the table in a bit more detail. All comparisons are based on identically sized members.

a. **Modulus of Rupture** — This refers to the bending strength of the wood, useful in considering spars and other structures that take the loads in bending. Observe that fir is 23 percent stronger than spruce under this heading.

b. **Modulus of Elasticity** — This refers to the material's stiffness. Observe that fir is a bit stiffer than spruce. An aircraft with a very high aspect ratio wing, such as are common to sailplanes, would ride a little less smoothly in rough air if it had a fir wing than if it had one of spruce; it wouldn't bend as much for the same load. However, pilots of the little, stubby-winged airplanes would hardly notice the difference since the wings on these aircraft are very stiff anyway.

c. **Compressive Strengths, Parallel and Perpendicular to Grain** —Observe that the compressive strength of fir, parallel to the grain, is 39 percent higher than spruce and the compressive strength perpendicular to the grain is higher by 55 percent. Aside from the obvious conclusion that a fir spar is stronger than a spruce spar of the same size there is an additional one; the bearing strengths of bolts in fir are higher by the same percentages. If you are concerned about bolts tearing out of a spruce member you can improve things by making the part of fir.

d. **Weight** — Since the table shows fir to be about 23 percent stronger than spruce (using the Modulus of Rupture as the basis for comparison) you would certainly expect it to be heavier. It is. In fact, it's more than 23 percent heavier — by 3 percent. It isn't as "efficient" as spruce on a pound for pound basis but it is very, very close.

If you're designing your own ship this small difference is quite meaningful in terms of saving money because, although fir is heavier than spruce it doesn't require **as much of it** to do the same job. Consequently the 3 percent value is good.

On the other hand, if you're building from plans you don't want to use "less of it." You certainly don't want to reduce the size of the spars and other critical structure so as to hold the weight increase to only 3 percent.

This is foolishness. Besides making the designer mad you will have all kinds of problems with this approach, trying to make things fit. You'd best leave everything just as it is and accept, as the table shows, an increased strength of 23 percent and an increased weight of 26 percent.

Obviously, although fir is 26 percent heavier than spruce your whole ship won't be that much heavier. You'll pick up that 26 percent only where you have used fir instead of spruce.

If your (spruce) ship weighs, say, 800 lbs. empty you have probably around 100 to 150 lbs. of wood in it. With fir your ship would pick up 25 to 35 lbs. more, which isn't very much, all things considered.

e. **Cost** — 150 lbs. of spruce (which is an interesting way to measure it, isn't it?) will cost at the very least $100.00, based on $1.50/board foot (the lowest price I've seen anywhere).

(Prices indicated in this article are based on pre-inflation prices for materials.)

TABLE 1
Some Comparisons, Spruce vs. Douglas Fir at 15% Moisture Content
(From ANC-18, 1951 Edition)

Characteristic	Spruce (psi)*	Fir (psi)*	Compared with Spruce, Fir is:
Static Bending			
(Fiber Stress at Prop. Limit, F_{bp}	5300	5900	10% stronger
(Modulus of Rupture, F_{bu}	9400	10900	16% stronger
(Modulus of Elasticity, E_L	1380x10^3	1480x10^3	7% stiffer
Compression Parallel to Grain			
(Fiber Stress at Prop. Limit, F_{cp}	3530	4220	19% stronger
(Maximum Crushing Strength, F_{cu}	4700	5600	18% stronger
Compressive Strength 90° to grain, $F_{cu}T$	740	1020	38% stronger
Shear Strength Parallel to Grain, F_{su}	990	950	4% weaker
Weight (w) lbs. per cu. ft.	28	33	18% heavier
Ratio, "strength" to wt. (F_{bu}/W)	335	330	1.5% "weaker"/lb.

*Unless otherwise specified

The same number of board feet of fir, figured at about 30 cents/board foot (fairly representative) will cost about $20.00, for a saving of $80.00. If you insist on using spruce anyway, in spite of the added cost, you will be paying $80.00 to save 35 lbs., or about $2.30 per lb., **plus** shipping costs.

If you build your 800 lb. airplane out of spruce, and if it is powered by a 65 hp engine which pulls it at 90 mph, 35 lbs. of added weight would reduce the top speed by about 2 mph. Your stubbornness in using spruce, or whatever you choose to call it, will cost you $40.00 per mph.

f. **Workability** — Fir is a little harder to work with than spruce. However, few woods are as nice to work with as spruce, so you'd expect it. Fir splits a little easier than spruce and since it is a harder material it doesn't sand quite as well. You also have to be a bit more careful in planing, watching the grain direction a little more attentively so you won't gouge out a piece by planing against the grain.

In conclusion, vertical grain Douglas fir is a fine material for use on aircraft. Although it is a shade heavier and stiffer than spruce and harder to work with, it is both cheaper and stronger — and you can get it. Use the same criteria in picking out the right pieces as you do with spruce; vertical grain, about 8 to 14 grains per in. and grain runout not to exceed 1 in. in 15. And use that $80.00 plus for other purchases you need for that dream ship of yours.

One of the most popular amphibians among homebuilders is the Osprey II. The all wood, two place amphibian is powered with a 150 hp engine and cruises at 130 mph. If operating in a salt water environment, the all wood structure eliminates corrosion problems that plague aluminum amphibians.

Characteristics, Sizes and Weights of Standard Plywoods

Physical Properties of 3-Ply Plywood

(8 to 10 percent moisture)

SPECIES	Average Specific Gravity, oven-dry weight, air-dry volume	Column Bending Modulus, lb. per sq. in., force applied parallel to grain	Modulus of Elasticity, column bending, 1,000 lb. per sq. in.	Tensile Strength, lb. per sq. in. parallel to grain	Tensile Strength, lb. per sq. in., perpendicular to grain	Compression Fiber Stress, lb. per sq. in.	Splitting Resistance, percent of birch	Service Suitable for A	B	C	D
HARDWOODS											
Ash, Commercial white	0.60	9,930	1,420	6,510	4,350	5,250	71				
Basswood	0.42	7,120	1,210	6,880	4,300	3,370	63				X
Birch, yellow	0.67	16,000	2,260	13,210	7,700	5,480	100	X	X	X	
Cottonwood	0.46	8,460	1,440	7,280	4,240	3,520	85				
Elm, American	0.52	8,680	1,220	5,860	3,990	5,180	75		X	X	
Gum, tupelo	0.50	7,760	1,300	6,260	3,760		60				X
Mahogany, African	0.52	8,070	1,260	5,370	3,770	4,280			X	X	
Mahogany, Khaya	0.48	8,500	1,940	6,390	3,780	4,880			X	X	
Maple, hard	0.68	15,600	2,110	10,190	6,530	5,620	114	X	X	X	
Maple, soft	0.57	11,540	1,750	8,180	5,380		106		X	X	
Oak, commercial white	0.64	10,490	1,340	6,730	4,200	4,950	85				
Poplar, yellow	0.50	8,860	1,920	7,390	4,720	3,750	51				X
Sweetgum	0.54	9,970	1,590	7,850	4,930		80				
Sycamore, American	0.56	11,040	1,630	8,030	5,220		77		X	X	
Walnut, black	0.59	12,660	1,740	8,250	5,260		77	X	X	X	
SOFTWOODS											
Cedar, Spanish	0.41	6,460	1,030	5,200	3,340	4,320	60				X
Cypress, bald	0.45	8,890	1,220	6,160	3,980	4,960	49				
Fir, Douglas	0.48	9,340	1,530	6,188	3,910	5,600	93				X
Hemlock, western	0.47	9,250	1,580	6,800	4,580		63				X
Pine, eastern white	0.42	10,130	1,570	5,720	3,340	3,840	31				X
long leaf						5,280					X
ponderosa						4,240					X
short leaf											X
sugar	0.42	8,050	1,310	5,430	3,690	3,680	47				X
Redwood	0.42	8,230	1,180	4,770	2,960		48				X
Spruce, Sitka	0.42	7,710	1,370	5,650	3,410	4,000	78				X

A — Hardness, abrasion resistence and fastening strength. B — Finish or appearance. C — For plywood that is to be steamed or soaked before being bent into a form or shape that is to be permanent after forming. D — For plat panels that require high bending or high column strength with minimum weight.

Sizes, Weights, and Stiffness of Aircraft Plywood

Thickness of Plywood, in.	Number of Plies	BIRCH Estimated Weight, lb. per sq. ft.	BIRCH Stiffness Factor (EI), lb. in. Long.	BIRCH Stiffness Factor (EI), lb. in. Trans.	POPLAR OR SPRUCE Estimated Weight, lb. per sq. ft.	POPLAR OR SPRUCE Stiffness Factor (EI), lb. in. Long.	POPLAR OR SPRUCE Stiffness Factor (EI), lb. in. Trans.
1/32	3	0.18	7.5	0.5	0.13	5.3	0.5
3/64	3	0.21	13.6	1.5	0.15	9.3	1.2
1/16	3	0.29	41	5.5	0.19	30.0	4.0
5/64	3	0.33	70	4.7	0.22	51.5	3.4
3/32	3	0.42	24	24.0	0.29	119	17.4
1/8	3	0.50	290	57.5	0.34	212	42
5/32	3	0.64	650	80	0.43	480	60
3/16	3	0.75	970	185	0.49	705	135
3/16	5	0.78	905	330	0.53	660	240
1/4	5	1.00	2,000	650	0.66	1,460	490
9/32	5	1.16	3,900	1,270	0.77	2,860	925
5/16	5	1.32	4,225	1,690	0.87	3,080	1,225
3/8	5	1.58	6,550	3,600	1.04	4,765	2,625
1/2	7	2.05	16,400	7,500	1.36	11,970	5,650
5/8	9	2.52	28,700	18,250	1.65	21,100	13,320
3/4	9	2.97	49,500	29,800	1.97	36,900	21,680
1	11	3.90	121,000	64,300	2.57	88,400	48,425

MATERIAL	STIFFNESS (EI) FACTOR[1]
Spruce (lightest commercial natural-density) plywood	416.0
Birch (heaviest commercial natural-density) plywood	178.0
Tegowood (high-density plywood) 1/45-in. birch veneers	30.0
Aluminum	22.0
Steel	3.1

[1] Obtained by multiplying the modulus of elasticity by the moment of inertia of a section 1 in. wide and of a thickness h equal to that of a sheet of the material 1 sq. ft. in area and weighing 0.43 lb. For birch, $h = 0.098$ in.; for spruce, $h = 0.153$ in.

Some Tips on Dry Rot in Wood

Because so many members of EAA are involved in rebuilding very old airplanes and in building new ones from materials bought from a variety of sources in and out of the aviation supply trade, some remarks about dry rot may be of interest and value.

The term 'dry rot" is confusing to many people since it implies, at first thought, a mysterious sort of rot taking place without moisture, most of us associating rot with dampness. Actually, "dry rot" cannot take place in the absence of moisture and the term probably has its origin in the fact that when wood affected by decay has been dried out completely from a damp state, the result is a crumbly material from which all the substances has been "dried out" by the rot.

What actually causes dry rot is a certain species of fungi, or tiny plants made up of microscopic threads invisible to the naked eye unless massed or matted together. The fungi which decay wood get their nourishment....for they are living things....from the cellulose of the wood. Fungi reproduce by bearing spores, which are microscopically small bodies something akin to seeds. These spores are carried about by the air and so it is that good wood can be infected by the accident of having a breeze bring some fungi spores to its surface. The fungi will grow in wood only if air and water are present, just like any other plant needs these two essentials of life. In addition, moderate temperatures are needed. One textbook says that the temperature range for various fungi is from 35 deg. F. to 100 deg. F., while another says that dry rot fungi requires a minimum temperature of 70 degrees to grow.

Lumber with a moisture content of less than 20 percent will not decay, which is why aircraft lumber specifications call for it to be kiln-dried to a moisture content of 15%; kiln temperatures kill off any fungi which may be present in new wood. If the wood is then kept dry by proper storage under cover and by suitable finishes when incorporated in a plane, it will not rot. Timber which is alternately wet and dry is slow to rot, by the way, since the spores can only flourish where moistures are present. Proper construction of wooden airplane parts requires some thought to preventing dry rot. Closed, completely unventilated spaces should be avoided. Protection is needed to keep rainwater and splashed water from entering structures insofar as in practicable, and drains must be provided at the lowest parts of all sections, enclosures, corners, depressions, etc. Any place in a wooden wing or fuselage which can collect and hold moisture is dangerous. Varnish or proper paint will seal wood against attack by spores but, naturally, if a pocket of water becomes trapped so it remains for a long time, it may soften the finish so that the wood is exposed. Once dry rot fungi begins growing in wood, it spreads rapidly and can jump from one piece of wood to another. If you cut out a rotted area that is no assurance that things will then be okay, because the rotted area may have left spores inside the plane which will start other rotted areas at any other damp place in the ship. Dry rot in wood is just as "catching" as colds are among humans. The use of anti-rot preservatives is common in the marine field but has to be approached with caution in airplane work because of possible adverse effect on glue joints and finish durability. Some woods are much more susceptible than others to dry rot. Birch, for example, is quickly attacked by dry rot, while some of the many species of Mahogany show a high degree or resistance. Cases are on record of fungi-susceptible inner plies of plywood dry-rotting out from between more resistant outer plies. Early traces of dry rot are slight whitish discolorations in the form of specks and streaks under the varnish not readily explainable as being caused by checked varnish. Advanced dry rot is indicated by dark brown or black discoloration and wood which crumbles under a knife point. Sound wood rings when rapped, rot-infested wood has a dead sound. There are many types of fungi and stains which attack wood and so if you are in doubt about the piece of lumber, study up on the whole subject on pages 107 thru 118 of ANC-19 bulletin "Wood Aircraft Inspection and Fabrication", from Gov't Printing Office, Washington 25 D.C. I once read a CAA report about a fatal accident in an old Waco 9 some years ago in which one of the lower wings came off during acrobatic flight. It seems that the woodwork of the catwalk at the wing root had become affected by dry rot which spread down to the spar butt and let the spar fitting pull loose under strain. If you are building a new plane, be damn sure your wood conforms to aircraft specifications, and if you are rebuilding an old-timer, look everywhere for the slightest suspicion of rot. The thing that makes dry rot such an insidious menace is that few people realize it is an infection of the wood, capable of developing in the most surprising places.

The Use of Non-Certified Wood

By W. Samuel Evans, EAA 18974

NONCERTIFIED WOOD should not be used for aircraft structural members subjected to tension either directly, as in longerons, or through bending, as in spars. For structural members subjected only to compression, select straight-grain commercial wood can be safely used.

Because of the high cost of certified aircraft spruce, it is not surprising that many homebuilders consider the substitution (1) (2) of select Douglas fir lumber at about one-quarter the cost. This writer took the Douglas fir approach in the design of the "Volksplane" until coming across a report (3) covering an apparently little known wood characteristic called "compression failures." As a result of this report and the investigation which followed, I found myself with considerable redesign in switching back to certified spruce for main structural members. However, in the process, some knowledge was gained which I feel should be passed on to the membership.

As a result of the above report, the Forest Products Laboratory was contacted with very good results. The cooperation of these people in the preparation of this article was very gratifying. In particular, I would like to thank Maxon Y. Pillow, in charge of Structure Property Relations Research, for his correspondence and telephone calls. Not only did he provide pertinent reports, but also examined samples that I submitted and provided other samples containing compression failures so that I might be able to recognize the condition.

Getting down to cases, the selection of any wood on the basis only of density, the slope of grain, and numbers of annual rings per inch overlooks the problem of compression failures. This is a condition of the wood fiber, due to excessive compression along the grain, which can develop in the standing tree during high wind storms. It can also be caused by felling over uneven ground, or even by rough handling of the timber in processing. Where this characteristic is present, the material involved will usually fail abruptly at relatively low stress levels and is therefore unsuitable for aircraft use. This is true principally for tension and bending members.

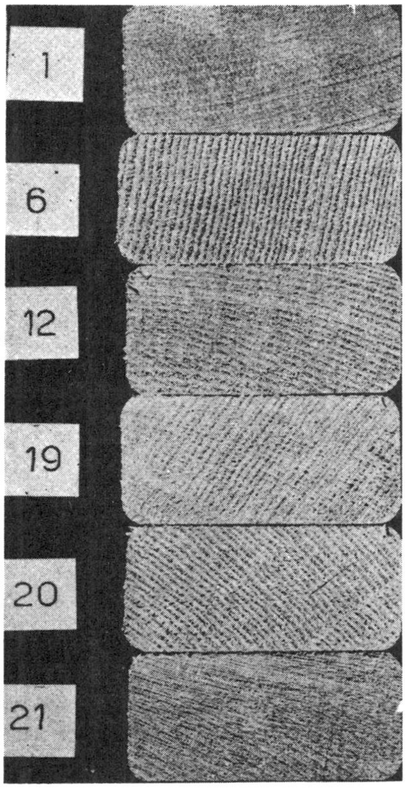

FIG. 1. Cross sections of ladder-rail samples that showed no fiber breakage.

FIG. 2. Four cross sections cut at about 2 ft. intervals from one end (13) and one from the opposite end (14) of a 16 ft. ladder rail, all showing fiber breakage caused by the presence of compression failures.

At this point, the reader may begin to question the probability of occurrence for this condition just as I did when I wrote FPL for statistical data. In return, a letter and report were received which covered a study made by FPL for ladder manufacturers who were having a lot of trouble with sudden breakage, during service, of apparently sound spruce ladder rails. According to the report (2), FPL went into the ladder shops, made a selection of stock, put this through a carefully documented series of tests and isolated the source of trouble as compression failures. Visual examinations also indicated the presence of this condition in an appreciable percentage of a large number of ladder rails in storage. The FPL ultimately set up an inspection procedure for the ladder manufacturers which included the training of personnel for recognizing this characteristic.

Now, if the ladder people and their insurance companies were concerned with this item, how much more should the homebuilder be concerned? So, what to do about it? Frankly, I don't believe that the average homebuilder will take the time to become proficient at detecting this condition and, for this reason, I feel that certified wood is the answer. Moreover, when you consider that even a minimum homebuilt will represent an investment of close to $1,000.00, not to men-

(Photographs courtesy of the U.S. Forest Products Laboratory)

FIG. 3. End grain surfaces of Sitka spruce lumber showing fiber breakage in the part below the dark line in comparison to wood free from this characteristic above the line.

tion the man hours involved, it is hard to justify any other approach. If the certified material is limited to tension and bending members, the added cost is small compared to the overall cost and is certainly cheap insurance.

Lest the reader get the impression that the purchase of certified spruce gets him home safe, let me hasten to add that it is also important for him to make his own inspection even with certified wood. This is because random inspection is the accepted rule with most products which means that not all items of a batch are tested, but just enough samples to assure a high probability of the entire lot being right. MIL-S-6073, which is the military specification for aircraft spruce, also follows this procedure. Section "F" of the specification calls for one piece out of every 20 pieces in a shipment to be selected at random for inspection and testing. If this piece meets the specification, all 20 are considered to be satisfactory. However, if the one piece fails to meet the specification, then all pieces in the shipment must be individually inspected and accepted or rejected on the basis of these tests.

Because of this procedure, aircraft companies have made it a policy to inspect each piece that they receive, and so should the homebuilder. The procedure is as follows:

With a plywood, or novelty cut-off saw blade capable of making a smooth cut, remove about 2 in. from each end of all boards and examine the end grain under a hand magnifying glass. With certified wood, the chances are that the fibers will look as in Fig. 1 and the upper part of Fig. 3. If, however, you see the condition shown in Fig. 2 on both ends of the board, don't use it for a tension or bending member. If only one end is bad, cut it off from that end in 2 ft. sections until it looks right. The remainder should then be good. In addition, check the planed surface carefully under suitable light for evidence of compression fracture across the board as shown in Fig. 4. If you see anything suspicious, apply a little light sanding and if that doesn't remove it you have a fracture.

Now, you might wonder why you can't do the same with noncertified wood. The answer is that you can, but I'm assuming that you will not become very expert at this, and consequently the odds will be in your favor with the certified material. What you are observing in Fig. 2 is not the compression fracture itself (this can usually be seen only under a high power microscope), but the breaking of the fiber due to the microscopic fracture. Sound fiber will shear clean when smoothly sawed, but fractured fiber will break away under saw tooth impact. It should be noted here that large or dull saw teeth can tear the sound fiber and give the appearance of compression failures, hence it is important to get a smooth cut.

FIG. 4. Large compression failures in spruce lumber that are visible as lines across the grain (slightly magnified).

In conclusion, a question arises regarding those designs that have been successful in using commercial fir throughout. You will note that in these cases the builder usually protects himself by going to a lot of laminating which pretty well limits any flaw present to a single lamina at any given cross section provided, of course, that the board isn't simply reassembled as sawed apart. Laminating involves a lot of labor, but unless you are willing to go that route, you had better use the certified material. To quote the people at FPL . . . "In our opinion, it is at least as desirable for an individual building an aircraft to have adequate inspection for wood characters affecting strength properties as should be provided for commercially built planes." Finally, for those homebuilders who are already well along on aircraft using commercial wood, the solution is either static testing which is time-consuming and costly, or very careful flight testing by a qualified pilot.

REFERENCES:

(1) "The Use of Douglas Fir as a Substitute for Spruce"; by Stan Hall in *SPORT AVIATION*, June, 1966.

(2) "Selection and Evaluation of Wood for Aircraft Use"; by Noel J. Becar in *SPORT AVIATION*, September, 1966.

(3) Forest Products Laboratory Report No. 1588 . . . "Detection of Compression Failure in Wood."

(4) Forest Products Laboratory Report No. 1733 . . . "Studies of Compression Failures and Their Detection in Ladder Rails."

Part 1

'Dry Rot' in Wooden Aircraft

By Clarence Sutton, EAA 4395

In a previous article I tried to answer some of the questions which come up when wooden aircraft are under construction. Inevitably, the subject of rot is mentioned, and as there is so much interest in all-wood lightplanes today a detailed discussion of rot seems in order. Nothing could be more short-lived and dangerous than a wooden airplane designed and built in ignorance of the cause and control of wood rot. On the other hand, if correct techniques are used a wooden plane can last almost forever.

It would be well in the beginning to explain the terms "rot" and "dry rot". The latter is in common use, and many people think that it must be something different from plain rot. This is not the case. Actually there is no such thing as "dry rot". For rot to develop, *there has got to be moisture.* If wood is kept dry, it will never rot. The term "dry rot" seems to go far back into nautical history; men working on ships would discover rotten places which were dry, by reason of the rot process having terminated, and concluded that they had found "dry rot". They usually found it above the waterline, and this heightened their impression that it was "dry rot". The truth is that what makes any boat rot out is fresh water seeping in from rain and condensation. Wood which is kept thoroughly waterlogged *will not rot.* The planking of a ship is the last part of it to rot, and even then the rot usually starts on the inside surfaces.

Wood rot is caused by fungus. This is a form of plant life which instead of growing in the ground and deriving nourishment from the soil, grows in wood, feeding on the cellulose content of the wood fibers. There are many forms of fungus which will attack wood. The seed of a fungus plant is a microscopic thing called a spore. The air is filled with spores and they settle out on every surface. You cannot cut a piece of wood and assemble it into an airplane without a goodly number of invisible spores settling onto it. Spores have a long dormant life, and one year, five years or ten years after being deposited on wood can come to life. They act much like any other seeds and will not grow in cold, dry places or in waterlogged, oxygenless places. When the air temperature goes up to a steady moderate temperature . . . about 70° F. for common types of fungus . . .and the wood's moisture content reaches a stage we would call "damp", life begins.

A germinating spore sends out microscopic, hairlike strands which might be likened to veins. They seek out the cellulose of the wood, absorb it, and feed it into the growing plant. This early stage of rot growth is insidious to the aviator, for it can appreciably weaken wood without changing its external appearance. Such wood suffers from brashness, or brittleness. As the fungus develops, the wood darkens, the microscopic hairs become so numerous as to form visible masses, (Fig. 1), and the wood begins to warp, shrink, and crack. In the final stage, the growing plant "fruits". Toadstools, puffballs, and the leathery growths on dead trees are all the fruiting bodies of mature fungus plants. The powder which issues from puffballs consists of spores, or seeds. A mushroom three inches in diameter may give off one to two billion spores - so you see how probable it is that all wood will bear spores!

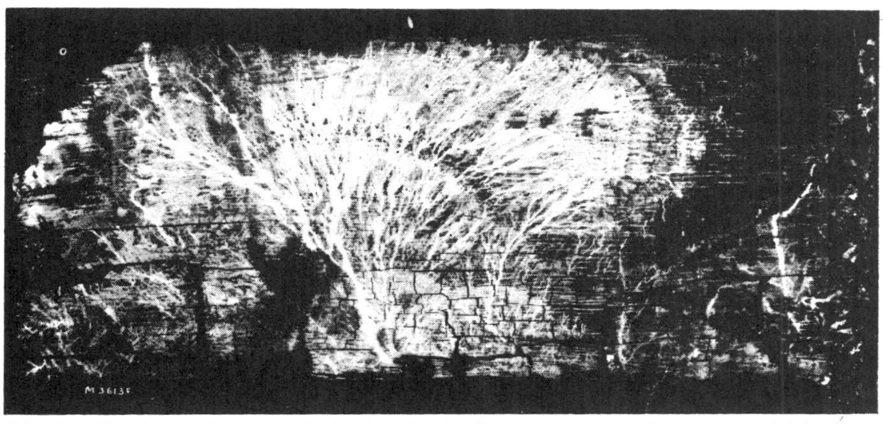

Fig. 1. Wood rots because it is the food of some forms of fungus, which send hairlike feelers throughout the cell structure to absorb the wood's cellulose. This is an example of wood decay in its last stages. Long before this state is reached, the feelers can do invisible damage by making the wood brash, or brittle. Fungus breeds only when dampness and warmth is present.

Most paints *will not prevent rot.* There is nothing in oil-base paints which will kill spores. You can paint or varnish over some wood and all that is accomplished is to seal so many spores to the surface. Whether or not rot develops depends on subsequent moisture and temperature conditions. In general, wood will not rot when its moisture content is below 20%. This is why proper drying and seasoning of aircraft lumber is so important. If green wood is used in a boat, airplane or house, and it is covered with paint, the sap is sealed in and will provide an excellent source of rot-producing moisture. The relatively thin pieces of wood in an airplane dry out much sooner than do the big timbers of a ship and so we don't have much cause to worry about green aircraft lumber . . .especially when it is bought from reputable firms.

Incidentally, it should be realized that not all forms of fungus cause rot. What we call mold or mildew are typical examples of harmless fungi. The lack of fungus-killing ability of common paints is attested to by the fact that paint often mildews in damp places . . . the vegetable oils in paint are palatable foods for mildew fungus.

We prevent rot in different ways. Woods vary appreciably in their rot resistance; it does not depend on any one characteristic such as hardness, but, it appears, on the chemical content of various species. Soft cedar has wonderful rot resistance but hard birch has very, very low resistance. Oak has excellent resistance in temperate climates but mediocre resistance in tropical or semitropical areas.

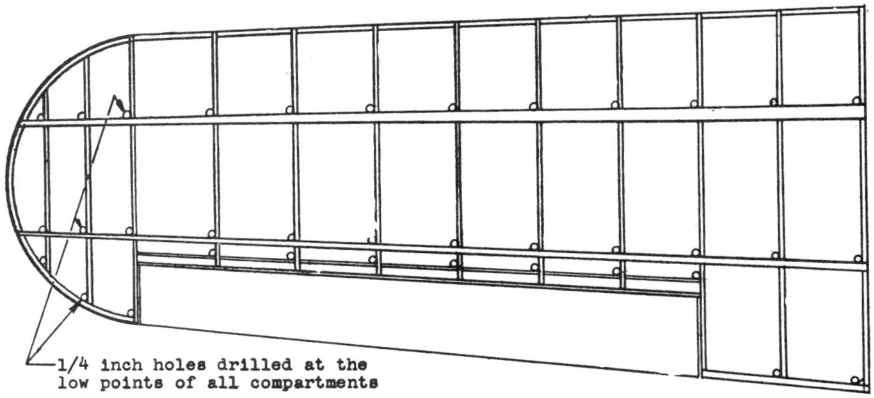

Drainage diagram of wing, direct method.

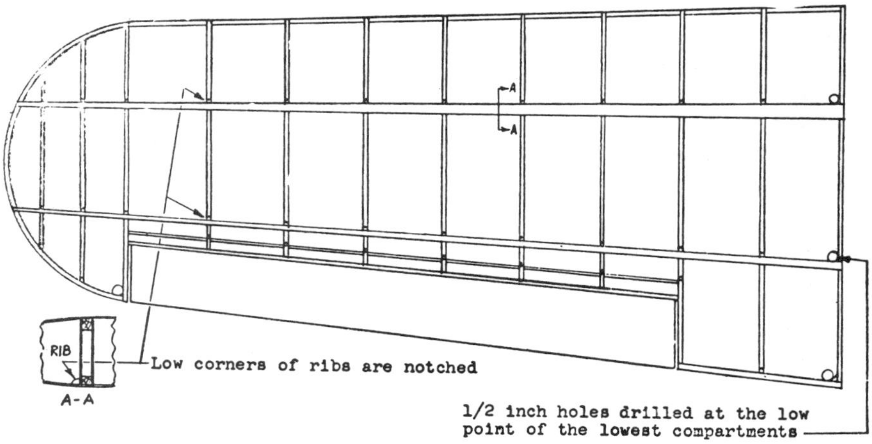

Fig. 2. The key to safe, rot-resistant wooden aircraft construction is to design the members to keep out water as much as is possible, and to let such water as does get in to get out again before it starts rot. The wing at top has drain holes bored at the lowest corners of all sections. The lower wing has notches in the structure so that all water will flow to the root and out the drain holes at the root.

Aircraft spruce rates as only fair in rot resistance. Here is a table of rot resistance for some common woods:

- Very Resistant - Greenheart, Black Locust, Teak, Osage Orange.
- Resistant - Most Cedars, Chestnut, Cypress, Honduras Mahogany, American White Oak, Redwood.
- Moderately Resistant - Douglas Fir, Hemlock, Hickory, Red Oak, Philippine Mahogany, Larch.
- Non-Resistant - Basswood, Cottonwood, Elm, Spruce, White Fir.
- Perishable - Ash, Balsa, Beech, Birch.

This rating is based on the average life of test specimens buried in the ground — which is a more severe rot-breeding condition than normally found in aircraft. The woods rated "very resistant" may last over 25 years; "resistant" 15 to 25 years; "moderately resistant" 10 to 15 years; "non-resistant" five to ten years; "Perishable" less than five years.

In addition, it is known that heartwood is more resistant to rot than sapwood, in properly seasoned lumber. Buying certified aircraft spruce from a reputable firm is our best guarantee that we're getting the safest, most durable wood. As was mentioned before, when moisture content exceeds 20%, most woods are susceptible to rot. Also, wood which is too wet is apt to be deficient in strength. A moisture content of not less than eight percent or more than 12 percent is recommended. Bone-dry wood is apt to split when bent. Moisture content can be checked by a process involving weighing a piece of wood, putting it in an oven, and re-weighing when dry. There are also electrical moisture content meters which are used in production work. Both are detailed on pages 54-56 of ANC - 19, "Wood Aircraft Inspection and Fabrication". Moisture inevitably gets through paint and varnish coatings and the moisture content of wood in an airplane will vary from week to week or season to season. Coatings merely slow down the rate of absorption and drying-out and help minimize checking or cracking.

Water gets into airplanes in various ways. Rainwater and water from melting snow trickles into open cockpits, into poorly-fitted doors, through stabilizer slots in the fuselage, through wing strut and aileron cable holes in the top surface of a wing, and so on. Ground water is splashed up into landing gear wells and is sucked into the wing and fuselage by the differential in internal and external air pressure. Condensation forms in closed or poorly ventilated places. On all-wood airplanes it is absolutely essential to provide drain holes at the lowest points of any and all cells or sections so that water will drain out promptly. Fig. 2 shows how a wooden cantilever wing can have either multiple drain holes leading to the outside, or a series of notches on internal corners which will let water flow to an outlet hole in the lowest place. When drain holes are used they must be located actually at the lowest points, and not an inch or two away from these areas. Otherwise small puddles will form. Very careful drilling is essential in order to get them at the low points, yet not accidentally drill into structural members. Quarter-inch holes are acceptable, but three-eighth inch holes are better as they are less apt to become clogged. The use of seaplane-type, cupped drain grommets is recommended, as they shield drain holes from flying water. The best test of all is to pour water into a structure before the top covering is applied, and study the effectiveness of the drain system. *Any pockets of water which may remain are dangerous rot-traps.* Enclosed places, such as the hollows in a box spar, should be ventilated with at least one air hole and preferably two to allow cross-flow.

Part 2

'Dry Rot' in Wooden Aircraft

By Clarence Sutton, EAA 4395

Much care is needed where metal fittings touch wood. Rust stains seem to destroy whatever rot resistance a wood may have, and if an unpainted fitting is bolted against bare wood, rot will take place sooner or later. Steel parts should be zinc chromated or cadmium plated, and the wood should be varnished. The DeHavilland people, who used wood a great deal in some of their Moth sportplanes and the Mosquito, used to set fittings into marine glue. If the varnish surface is broken when bolts are drawn up, or develops hairline cracks as it dries with age, some protection is lost. This could be fatal if located in an area subject to moisture accumulations. If adequate drainage and ventilation is provided, it could be of little consequence.

Modern practice is to make effective use of rot-inhibiting chemicals. We have always done this in one way or another on boats and on the ground. Creosote is an excellent fungicide for fence posts, railroad ties, etc., for it is cheap and long-lasting. However, it has a strong odor and causes paint to peel, so it has limited uses in applications where people will live or work. Salt and brine will kill fungus just as effectively as it will kill lawn grass, and old time wooden ships used to have rock salt packed into rot-susceptible crevices. Today we have quite effective wood preservatives, such as copper napthenate and, especially, pentachlorophenol. The latter can be mixed with turpentine or other non-oil vehicles and when applied to millwork, boats or aircraft it impregnates the wood with a highly effective preservative that is colorless, odorless and which will take paint and varnish well. Good grades of wood preservative are made by many firms; the can should have a listing of ingredients and if pentachlorophenol and a non-oil-base vehicle are indicated, any brand under consideration should work well. Apply the fluid copiously, either by immersing small parts or brushing it on large areas. **Do all drilling and gluing first.** Never try to glue surfaces which have been treated with preservative. Wipe off all oozing glue with a rag and warm water during assembly work — if drops of glue are left on woodwork, they will keep preservative off of those areas and later when the glue dries hard and cracks off, unprotected areas will be created. When I was rebuilding my antique Monocoupe 90A I found such places, and the basswood was black with mold!

Preservatives can in time leach out of the wood fibers. This is a big problem in boats and houses, and in airplanes it depends on conditions. A good four-or-five-coat varnish job on aircraft woodwork is highly recommended as it seals preservative in as well as controlling excessive drying and soaking of the wood as the weather changes.

EDITOR'S NOTE: EAAers interested in wooden aircraft should not fail to study two fine government books on wooden aircraft. They were written during World War II when materials shortages forced designers unfamiliar with wood to use it, and have been reprinted in up-to-date editions since then. They are "Wood Aircraft Inspection and Fabrication", Bulletin ANC-19, and "Design of Wood Aircraft Structures", Bulletin ANC-18. They can be ordered from the Superintendent of Documents, Government Printing Office, Washington 25, D. C. Send a postcard there first to determine present availability and cost.

The solvents used in nitrate and butyrate dopes will soften varnish, so a fabric-covered wooden airplane structure should have dope-proof paint applied where fabric touches wooden members. Another method is to press aluminum foil strips onto the last coat of varnish while it is still tacky. Use household aluminum foil cut into strips with a cutter made of razor blades. The Navy and many commercial manufacturers have used this trick with success.

Never use plain iron or steel nails and tacks in aircraft work. They will rust, and lead to rotting of the surrounding wood. Cement-coated nails are all right, as are cadmium-plated and copper tacks.

Smooth, neat workmanship aids materially in rot prevention. Rough saw cuts leave a surface with a host of ridges and valleys which, even though they may be varnished, provide holding places for deposits of dust and silt. In wet weather such places hold moisture and breed rot. A smooth surface is much to be desired. In Japan during the occupation I found Japanese trainers in which the cedar spars and ribs were finished as smoothly as parlor furniture. In the vicinity of the cockpit, cabin doors or wing and tail surface slots, pockets formed by plywood gussets on the lower longerons should be filled with wooden blocks, preferably beveled on their top surfaces, to keep water from collecting during every rainstorm.

Modern wood preservatives give increased safety in wooden airplanes by providing a last-ditch barrier to rot formation. One of these samples was treated, the other was not, and both were exposed to rot-breeding conditions for equal periods.

Many of us work with antique airplanes as well as new homebuilts, and here again a knowledge of wood rot principles is vital. Aircraft woods stain and discolor from various causes. There are harmless forms of mildew which can turn wood black. There are also harmless fungus stains which turn it blue, gray or black. Chemical interactions between glues, finishes, metal parts and fuels can also cause discoloration. The author has inspected many war-surplus wooden planes and feels that any chipped or cracked area in paint or varnish should be suspected. Often plywood wings are weather-checked on their top surfaces and grayed wood fibers are visible. Since the sun usually dries water out of such cracks quickly, rot does not too often get started in such places. However, if the plies of the wood seem to have separated such as to form pockets or crevices into which water could be drawn by capillary action, rot could be well established. Any plywood which shows signs of delamination for more than a small fraction of an inch in from its edge, or from the edge of check lines, should be viewed with doubt.

Often a mechanic will be sorely taxed to decide if discolored wood is rotten or merely stained. If a high-power pocket magnifying glass or microscope is available and traces of fine, hairlike growths can be seen, the wood is obviously infected with rot and is to be condemned. Another test is made with a very sharp-pointed knife such as an X-Acto modelling tool. Insert the tip diagonally into the wood and try to lift a sliver up off the surface. If in contrast to an obviously sound section, the suspected area produces short, crumbly, easily-lifted chunks instead of long, hard slivers, rot is probably present. On end grain, such as the butt ends of spars, rot may be indicated if scraping produces punky, powdered wood.

Spar butts are located near holes in the wing covering where fittings project, and are a common and dangerous breeding place for rot, especially in airplanes stored out-of-doors. Wood inevitably swells and shrinks with weather changes and this often ruptures the surface of varnish or sealer on spar butts. If end grain cracks are produced, water can get in and as ventilation is often poor, rot begins. Rot spreads much faster with the grain than across the grain, and may soon reach the bolt holes of the fittings. If spar-butt cracks are not over one-sixteenth in. deep, file the butt smooth with a cabinet file, removing that amount of wood. Then cut a piece of one-sixteenth in. aircraft plywood to make a cap for the butt and attach it with ¼ x 20 cement-coated nails and glue. This protection can also be incorporated on new airplanes.

A wooden airframe which has been designed to shed water, let water drain, let humidity ventilate away, and which has in addition been given the protection of wood preservative chemicals and good surface finishes, can very easily outlast any steel tube airplane. In Europe there are airplanes with wooden fuselages such as the Aeromarine-Klemm, Avro Avian and Gipsy Moth which are thirty years old and still airworthy. With vastly improved modern glues that resist fungus and dampness, nobody needs doubt the safety and durability of wooden airframes.

Typical Wing Compression Rib Repairs

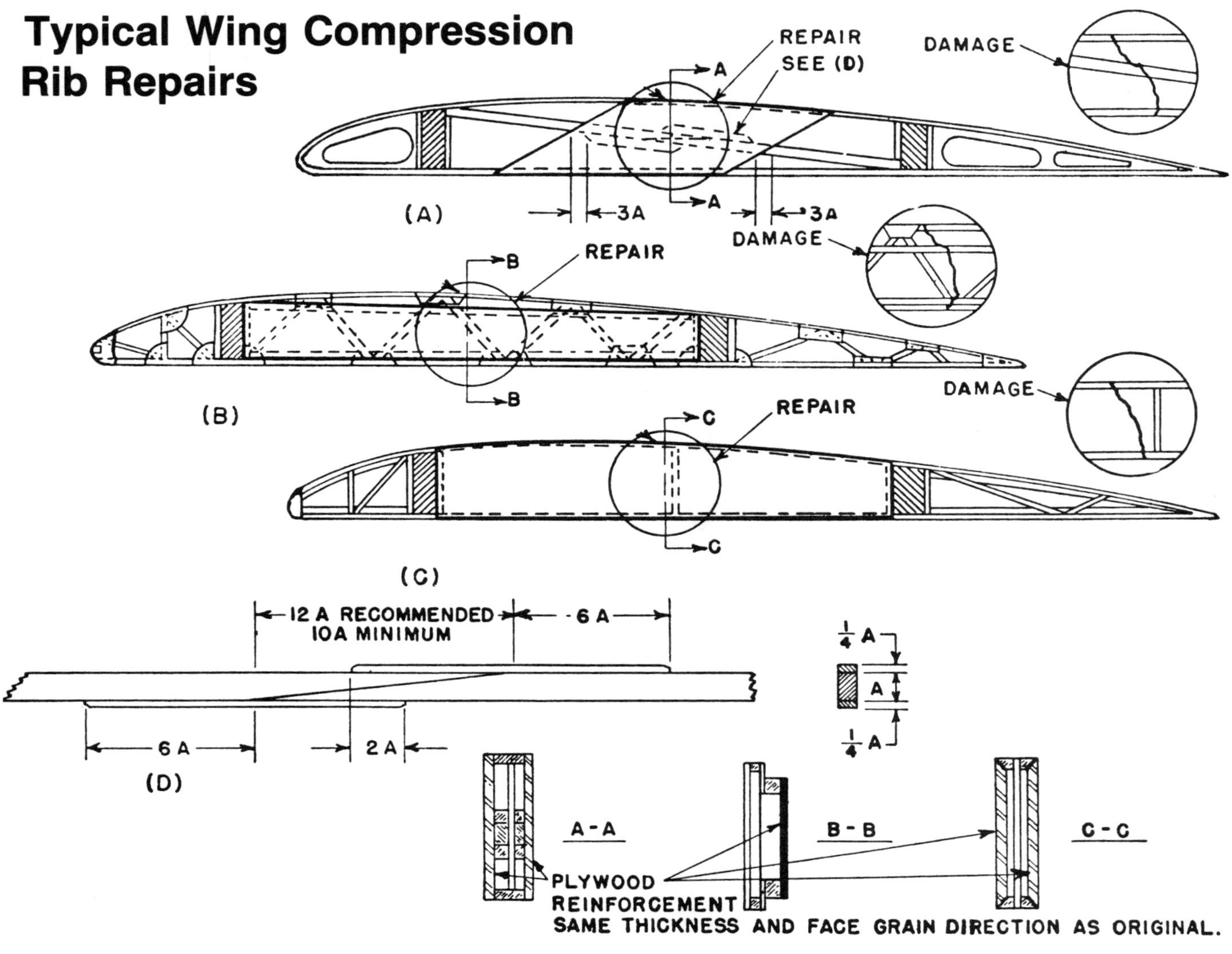

Wood for Light Aircraft

By Georges Jacquemin, EAA 3618

An amateur aircraft builder should have sufficient knowledge of the structure and properties of wood which will enable him to properly select appropriate materials for his aircraft. The major part of the structure of most light aircraft is made of spruce. Hard woods, such as birch, beech, ash, etc., are used mostly for blocks.

After a log has been sawn into quarters it is further cut into suitable sized planks. The pith, wane and some of the sap wood is cut away. Fig. 1 shows a cross-section of a log with two different manners of cutting planks. The lower quarter shows the most common method of cutting aircraft timber. The upper quarter shows a simpler method of cutting which gives only about half of the wood suitable for aircraft construction. Plank A is rift-sawn and most suitable for aircraft use. Plank B is slashed or tangentially-sawn and plank C is quarter-sawn.

After the planks have been cut, they are placed in stacks to prevent stains and mould and allowed to season for about 60 days. The total weight of water in freshly felled timber may amount to more than 100% of the dry weight of the wood substance. Moisture is present in the wood in two different forms — as free water in the cell cavities and as absorbed water in the substance of the cell walls. Most of the free water is evaporated at the end of the seasoning period while the cell walls are still practically saturated. This point is called the "fiber saturation point". During the seasoning period the timber loses weight but no appreciable deformation occurs.

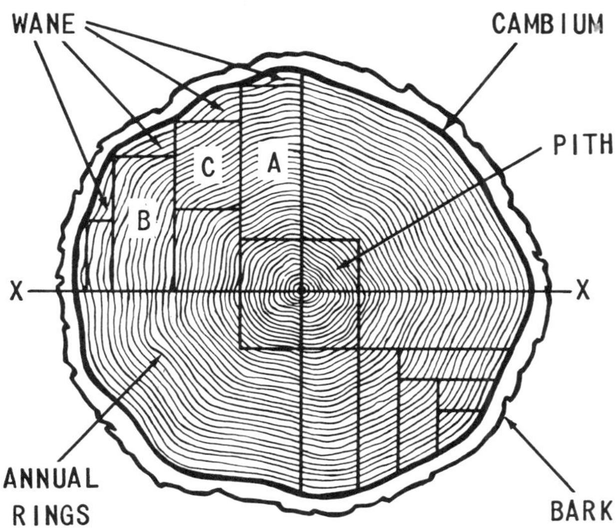

SECTION OF LOG SHOWING FIRST CUT

Fig. 1

When the wood reaches the fiber saturation point its moisture content is approximately 27%. This must be further reduced to about 14% to make the wood suitable for aircraft use. In contrast, 8 to 9% is recommended for furniture making. As drying proceeds beyond the fiber saturation point, the moisture contained in the cell walls begins to evaporate and the wood as a whole begins to show appreciable shrinkage which increases while the wood loses moisture.

Changes in the moisture content of timber affects its strength. Seasoning increases the hardness and stiffness of the timber but reduces its toughness or resistance to shock. When wood shrinks and swells the dimensional changes are not the same in all directions of the grain. Wood fibers do not shrink appreciably in length when they are dried, but they do so crosswise. Fig. 2 shows how different cut planks will shrink.

There is no practical method by which the amateur can find out the percentage of moisture in the wood he wishes to buy. In general wood sold as aircraft timber and delivered with a release note will have the right amount of moisture. Wood purchased without a release note or from the local lumber yard can be quite far from the required moisture content. Electrically operated moisture

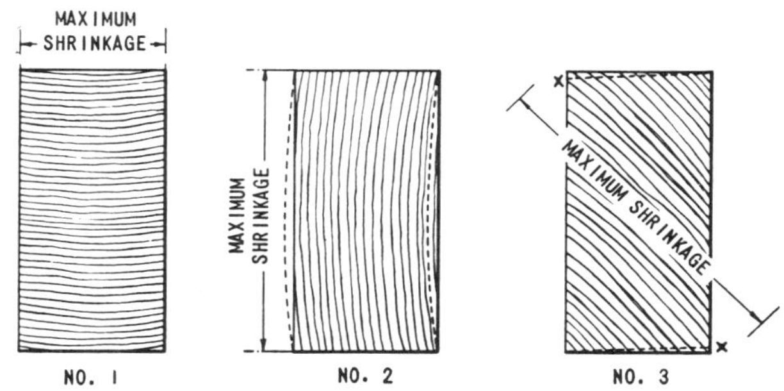

(DISTORTION DUE TO SHRINKAGE IS INDICATED AT XX)

Fig. 2

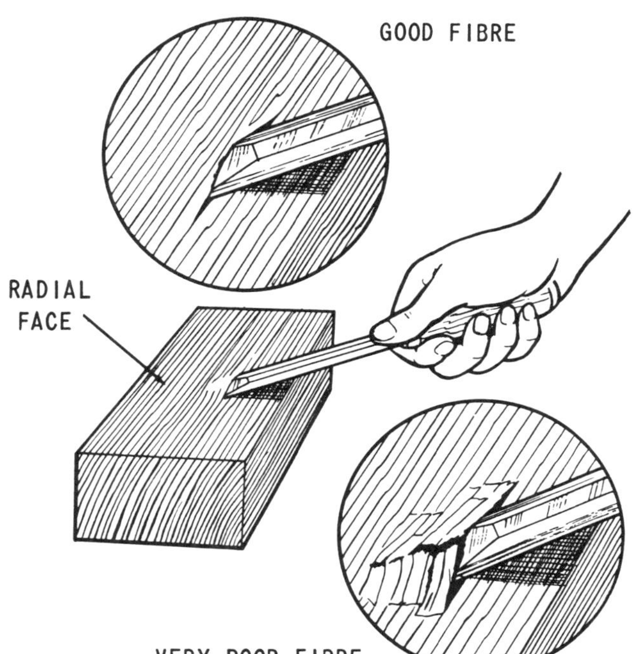

Fig. 3. Testing the wood fiber.

meters can be used. Fig. 3 shows a method of testing the wood fiber.

It can be seen from Fig. 2 why slashed planks are not desirable for aircraft construction. It would be impossible to obtain a stable flat plank from wood with grain as shown in Section No. 2. Splash planks can be used to cut small strips but rift-sawn planks must be used for all elements of the primary structure.

Timber is a complex organic structure and in common with most other natural products shows lack of uniformity. It is neither isotropic nor homogeneous. The weight of spruce varies between 20 and 36 lbs. per cu. ft. so that no fixed figure can be given for its specific gravity, but usually the average density may be taken at 27.5 lbs. per cu. ft.

The variation in density is not only due to moisture content but also to the rate of growth of the wood. A slow rate of growth will produce a heavier wood. Wood is heavier near the foot of the trunk of the tree and lighter at the top, so that the center of gravity of a plank is not exactly at the half point of its length.

The strength of spruce varies also within large limits. For stress analysis purposes it is usually taken at 8300 PSI tensiles strength parallel to the grain. The number of annual rings varies widely between specimens - anywhere from two to 30 rings to the inch. The best specimens average from 10 to 20 rings to the inch and a minimum of 6 rings to the inch is acceptable. The number of rings to the inch is related to the rate of growth of the tree. A large number of rings mean slow growth, higher density and the timber may be brashy and brittle. (See Fig. 6).

Wood can be classified into three grades with the following characteristics:

Grade A — Used for stressed parts. Minimum density 24 lbs./cu. ft. Moisture content: 10 to 17% Grain slope not less than 1 in. in 15 in., No knots

Grade B — Used for secondary parts. Density between 22 and 24 lbs./cu. ft. Moisture Content: 10 to 17%, Grain slope not less than 1 in. in 12 in., Pin knots may be present

Grade C — Not used for aircraft construction.

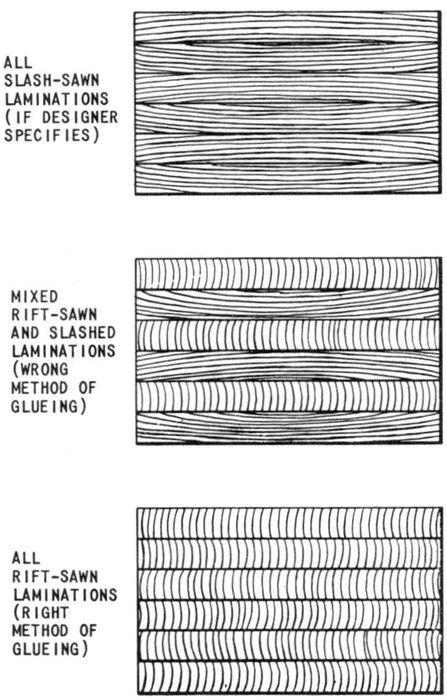

Fig. 4

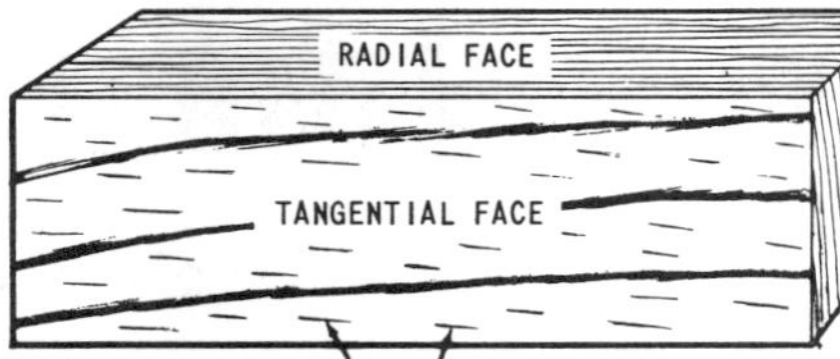

Fig. 5

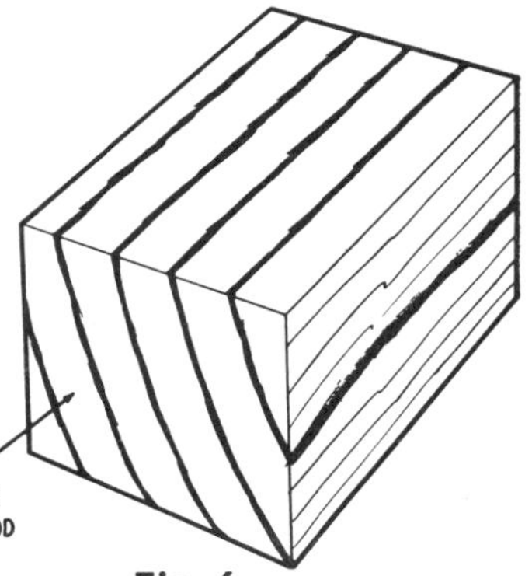

Fig. 6

It is generally understood that all lumber should preferably be rift-sawn. Fig. 4 shows how laminated parts should be glued. All sash sawn lamination is sometimes specified by the designer, otherwise rift-sawn lamination is mandatory. After gluing a number of laminations with casein glue, the part should be allowed to stand for at least seven days before finishing to final dimensions so that all excess moisture has time to disappear and the part will remain dimensionally stable.

The worst of all defects in wood is "dry rot", which is contagious. It is caused by a fungi which thrives in timber containing 20 to 22% moisture. Any signs of dry rot will reveal themselves in the shape of brown and yellow patches which crumble to powder at the slightest touch. The wood should be inspected on all its faces including the ends. If any signs of dry rot are present, it should be scrapped.

The inclination of the grain in wood for aircraft use should not exceed 1 in. in 15 in. for Grade A and 1 in. in 12 in. for Grade B. To find the inclination of the grain it is necessary to locate the resin ducts. They are apparent on the flower face of the wood as shown in Fig. 5. Their inclination can be measured and checked against the specifications. This check should be carried out on the whole length of the wood. It is important to find these ducts as they are an infallible indication of the correct inclination of the fibers. The fibers run parallel to the ducts.

There is no objection to using members with small pin knots provided that they are in the solid part of the wood (Grade B timber only). However, gum pockets, outward bulges, large knots are all causes for rejection of the wood for aircraft use.

Birch is the most commonly used of the hardwoods in North America, while beech and ash are used in Europe. In light aircraft construction only small parts, such as blocks for attachment of metal fittings, are made of hard wood. Although hard wood purchased from local lumber yards is slash cut, it is suitable for this purpose. The same remarks regarding straightness of the grain, moisture content and presence of defects as listed above for spruce also apply for hard wood. The number of annual rings per inch is usually lower than spruce - 4 to 10 rings per inch for ash. For walnut and mahogany the inclination of the grain should not exceed 1 in. in 12 in., and for ash 1 in. in 10 in.

Selection and Evaluation of Wood

By Noel J. Becar, EAA 725

THE PURPOSE of this report is to concentrate attention on the factors which need to be considered in the use of wood as a constructional material in the building of aircraft. This includes the consideration of suitable wood other than spruce, such as Douglas fir which, due to its ready availability and price, seems to be an ideal material for substitution, particularly in those areas where spruce is difficult or expensive to obtain.

All positive statements herein are based on material obtained from the following references which have been used in summarizing the data applied in this report:

NO. 1 ANC-18, DESIGN OF WOOD AIRCRAFT STRUCTURES (1951)

NO. 2 ANC-19, WOOD AIRCRAFT INSPECTION AND FABRICATION (1951)

NO. 3 AIRCRAFT MATERIALS AND PROCESSES, by G. F. Titterton (1951)

The Composition of Wood

As opposed to metals, which have identical composition and strength in all directions, wood can be considered as consisting of hollow fibers or tubes of indefinite length. These are firmly welded together by a cellulose cement which binds the fibers so firmly that when separation occurs, it is usually in the fiber walls rather than in the bond between fibers. This composition accounts for greater strength parallel to the grain than across it, both in tension and compression. In practically all wood, the light-colored layer next to the bark is called "sapwood", while the central portion which is darker is called the "heartwood." Heartwood is not fundamentally weaker or stronger than sapwood. After cutting, the heartwood is usually more resistant to decay, stain and mold, but the sapwood is more porous and pliable and therefore preferable where severe bending is encountered, such as in wing spars.

Wood swells as it absorbs moisture and shrinks as it loses it. Wood shrinks most across the grain, tangent to the annual rings, thus causing it to twist, cup, sliver and check. It shrinks one-half to two-thirds as much across the rings in a radial direction, thus maintaining spar dimensions more adequately. It shrinks very little parallel to the grain, or lengthwise; see Fig. 1. Plain-sawed or flat-grain cuts have the annual rings less than 45 deg. to the

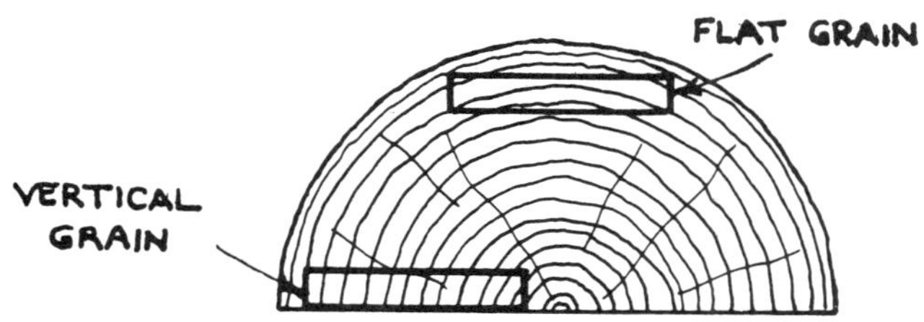

Fig. 1

wider surface. Quarter-sawed or vertical-grain cuts have the annual rings 45 deg. or MORE to the wider surface. All wood loses or absorbs moisture from the air until the amount in the wood balances the amount in the atmosphere. Certain coatings, such as varnish or paint, will reduce the rate at which it gives off or takes on moisture, but will not prevent it over a period of time.

Lumber stabilizes at a moisture content of from 12 to 15 percent of the dry weight of the wood. As the moisture content reduces, wood becomes stronger in all respects and tougher or more shock-resistant. For example, Sitka spruce becomes 3.9 percent stronger in bending for each 1 percent decrease in moisture content. The modulus of rupture, or maximum strength in bending, for Sitka spruce is 9400 psi at a moisture content of 15 percent, whereas it increases to 14,600 psi under bone-dry conditions. The maximum crushing strength parallel to the grain also increases from 5500 psi to a little over 10,000 psi under the same conditions.

On the other hand, increasing the moisture content to 27 percent decreases the maximum strength in bending to 6200 psi, and the crushing strength to approximately 3000 psi. This is mentioned in order to point up the commonly incorrect belief that wood should not be allowed to get too dry, whereas the fact is that one should be more concerned about any excess moisture absorbed by aircraft grade lumber. Do not make the mistake, however, of attempting to assign a higher strength value to lumber simply because it happens to be drier than the 12 to 15 percent normal value. As stated before, wood will stabilize at a moisture content dependent upon its surroundings, regardless of any protective coatings.

Some Spruce Substitutions

Let's consider some of the possible substitutes for aircraft grade spruce and see how the physical specifications compare:

Kind of wood	Wt. per cu. ft.	Strength in bending
Sitka spruce	28 lbs.	9,400 psi
Red pine	33 lbs.	10,800 psi
Douglas fir	33 lbs.	10,900 psi
Western hemlock	30 lbs.	11,000 psi
Port Orford cedar	29 lbs.	10,200 psi
Western larch	37 lbs.	11,000 psi
California red fir	28 lbs.	9,400 psi

It should be obvious from the foregoing that Douglas fir and Port Orford cedar are both capable of withstanding more stress, or load, than Sitka spruce per cubic inch, at a very small increase in weight. Let's examine the substitution of Douglas fir for Sitka spruce in a specific case, such as the front spar for a "Baby Ace." This spar is ¾ in. wide, 5⅛ in. high, and 149½ in. long, representing .332 cu. ft. of lumber. A spruce spar at 28 lbs./cu. ft. would weigh 9.3 lbs., whereas a Douglas fir spar at 33 lbs./cu. ft. would weigh 10.4 lbs. This is a 1.1 lb. increase per spar or an approximate increase of 4.4 lbs. for all four spars required in the "Baby Ace" wing.

When one considers, for an average, a local selling price of 60 to 70 cents per board foot for spruce and then contemplates the comparison of 30 to 40 cents per board foot for Douglas fir from the same dealer, 4.4 lbs. doesn't seem such a penalty weight-wise for such a saving in money. This is a comparison for the same-size spars, but if one were to take full advantage of the increased strength of Douglas fir as compared to Sitka spruce, it should be examined as follows:

First, compute the Moment of Inertia (I) of the "Baby Ace" spar:

$$I = \frac{w \times h^3}{12} = \frac{.75 \times (5.125)^3}{12} = 8.41$$

Where: w = width of spar
h = height of spar

If we substitute the allowable value of 9400 psi for spruce, representing the bending stress, or modulus of rupture (f_b), we can determine the maximum load this spar is capable of carrying.

$My = f_b \times I = 9400 \times 8.41 = 79{,}054$

Where: M = bending moment in inch pounds.
y = distance from neutral axis of spar to outer surface, on compression side

Note that the above answer obtained is the bending moment **times** the distance from the neutral axis of the spar to the outer fibers on the compression side. This is sufficient for our purpose, but actual bending moment could be obtained, if desired, by dividing the answer by the value of "y", or 2.5625.

By rearranging the bending stress formula, and substituting 10,900 psi for f_b, we can find the corrected value of I for a spar made of Douglas fir:

$I = My/f_b = 79{,}054/10{,}900 = 7.25$

Now we can solve for the reduced width of spar which will possess the same strength, made of Douglas fir, as the original ¾ in. wide spruce spar:

$w = I \times 12/h^3$ or, $7.25 \times 12/134.61 = .646$, or approximately 21/32 in.

The total volume of a Douglas fir spar as figured above, with its width reduced from ¾ in. to 21/32 in. would equal .291 cu. ft. Therefore, at 33 lbs./cu. ft. for Douglas fir, this spar would weigh 9.60 lbs. Compared to 9.30 lbs. for spruce, this represents a weight penalty of .30 lb. per spar, or only 1.2 lbs. for the four spars used in the entire wing!

To anyone who is familiar with stress computations, we wish to add that we realize the above is an over-simplified approach to the problem, but it **is** practical and simple. At the same time, it will not introduce any more error in the final analysis than is provided by practical deviation in wood properties and dimensions for this size and stress limit of spar.

Basic Requirements for Aircraft-Grade Lumber

We will summarize the requirements first, as a quick check-list for anyone attempting to hand-pick aircraft-grade lumber at a lumber yard. This is strongly recommended, as opposed to placing an order and taking whatever the dealer sends, as we have found that most reputable dealers do not object to a person hand-selecting their lumber.

Minimum number of annual rings per inch:

Sitka spruce	6
Red pine	6
Douglas fir	8
Western hemlock	6
Port Orford cedar	8
Western larch	8
California red fir	8

The following specifications are simplified as much as possible and apply to SOLID spar material only. LAMINATED spar requirements will be specified separately, as the majority of spars used in homebuilt aircraft are solid:

Maximum slope of grain:

In general:

No steeper slope than 1 in. in 15 in. with respect to the longitudinal axis of the board.

For solid spars:

Outer eighth of spar height must not slope steeper than 1 in. in 15 in.;

Adjacent eighth of spar height can deviate from above but shall not slope steeper than 1 in. in 10 in.;

Middle half of spar height can be as steep as 1 in. in 10 in.;

Spars must be edge-grained at least two-thirds the height of both vertical surfaces.

If diagonal or spiral grain, the **effective** slope must be determined.

This is equal to:

$$\sqrt{(\text{edge slope})^2 + (\text{side slope})^2}$$

Knots

The SIZE of a knot (see Fig. 2) means the distance between lines enclosing the knot and parallel to the edges of the face on which it appears.

The DIAMETER of a knot is the **minimum** distance between parallel lines (in any direction) enclosing the knot.

In general:

No knot shall exceed ½ in. in size **or** diameter.

For solid spars:

Within outer quarter of spar height, no knot is to be over 1/16 W in **size,** where "W" is the width of the spar;

Within middle half of spar height, no knot is to be over ½ W in **diameter.**

Pitch or Bark Pockets

For solid spars:

Not deeper than ⅛ W, nor wider than ¼ in. or ⅛ W, whichever is the lesser, and no longer than 2 in. or four times the distance to the spar corner, whichever is the lesser;

Distance between two pockets on the same face of the spar to be not less than six times the length of the shorter pocket;

For pockets in the same growth layer, distance be-

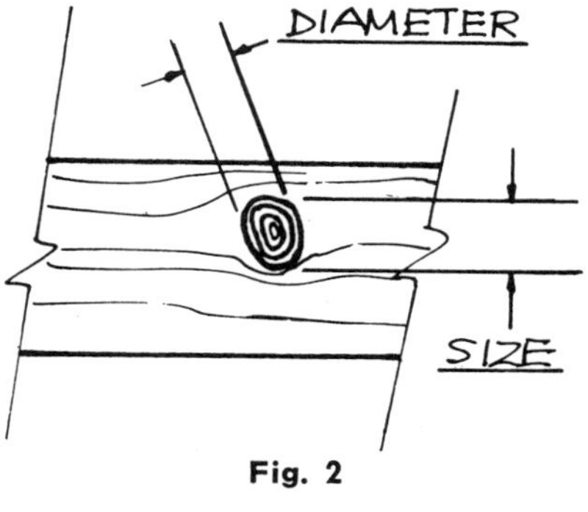

Fig. 2

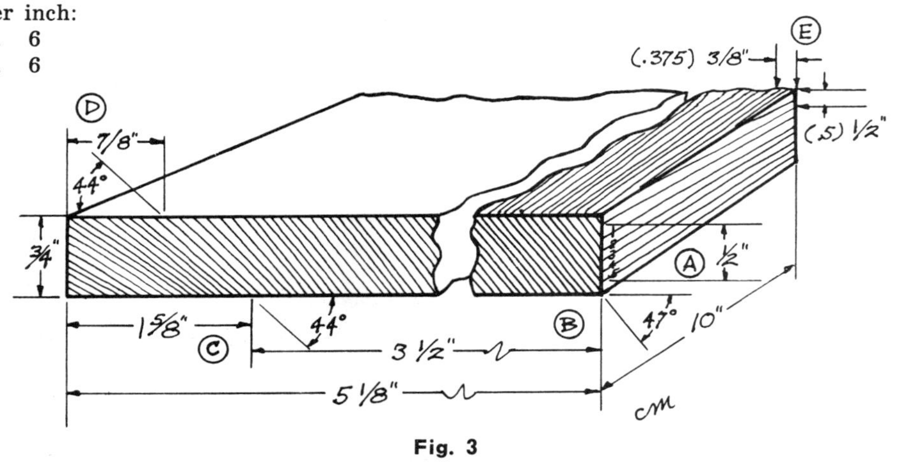

Fig. 3

tween pockets not less than six times the length of the longer pocket;

No pitch or bark pockets permitted in members less than 1 in. in either width or height.

Most lumber inspected would be covered by the foregoing items, but a few other items should be checked for, where applicable:

Regarding **slope of grain,** where a spar tapers in height, the grain slope should be measured relative to the tension side, or bottom edge as normally situated in a wing;

Regarding **multiple knots,** the sum of the **sizes** of all knots (both on the spar edge and adjacent quarters of the vertical faces), within a distance equal to five times the spar width, should not exceed ⅛ W, and the sum of the knot **sizes** within a length equal to the spar width should not exceed 1/16 W.

In the middle half of the spar height, the sum of the **diameters** of all knots on one face within a distance equal to five times W should not exceed ½ W.

If a knot is under 1/16 in. in **size** or **diameter,** it can be disregarded as an individual knot, but should be included in the limitations for multiple knots.

When the same knot appears on opposite sides of a spar, the average of the measurements on the two sides can be used for **size** and/or **diameter.**

If two or more knots are so close as to form a cluster around which the grain is deflected as a unit, the cluster shall be considered as an individual knot.

Summation of Requirements by Example

We realize that the application of the above requirements can be confusing, especially to one who has never had the need to apply them before. We propose, therefore, to consider the use of a plank of Douglas fir, assuming it will replace a Sitka spruce spar in a "Baby Ace" wing. We shall apply all the criteria necessary in order to determine if it will meet minimum requirements.

Fig. 3 represents the **least** favorable end of a certain plank we have selected for examination. The first item we shall check is the minimum number of annual rings per inch. Examining the widest spacing between rings that we can find in this plank at location (A), we find five annual rings in ½ in., or 10 rings per in. This meets the minimum requirement of 8 rings per in. for Douglas fir.

Checking next for vertical grain, we find at location (B) that the angle between the grain and widest side of the board is 47 deg. A slope of 45 deg. or more, by definition, establishes this as vertical grain. However, at location (C), we find the slope has changed to 44 deg., which defines this portion out to the left-hand end as flat-grain. However, for solid spars, we must meet the requirement that two-thirds of the spar height, on both surfaces of the spar, must be vertical grain. As 3½ in. is slightly over two-thirds of the height of a 5⅛ in. spar, and as the flat-grain condition only exists for ⅞ in. at location (D) on the opposite side of the spar, this condition will meet minimum requirements.

To examine for the maximum allowable slope of grain condition, beginning in the top right-hand corner at location (E), we note the grain runs out ⅜ in. in 10 in., or a slope of 1 in 26, (10/.375). However, the same annual ring runs out on the edge of the spar to the extent of ½ in. in 10 in., or a slope of 1 in 20, (10/.5). This indicates a spiral grain condition, so we must determine the **actual** slope from formula. A slope of 1 in 26 = 1/26 = .038, and a slope of 1 in 20 = 1/20 = .05, hence $\sqrt{(.038)^2 + (.05)^2} = \sqrt{.0039} = .062$, and $1/.062 = 16$, or an actual slope of 1 in 16, thereby meeting the requirement of a maximum slope of 1 in 15.

In ending this report, I wish to emphasize that the spar material pictured in Fig. 3 is definitely not a desirable specimen. It was presented simply to show material which is close to the absolute minimum that would pass as aircraft spar material per the FAA requirements, and still possess the required strength.

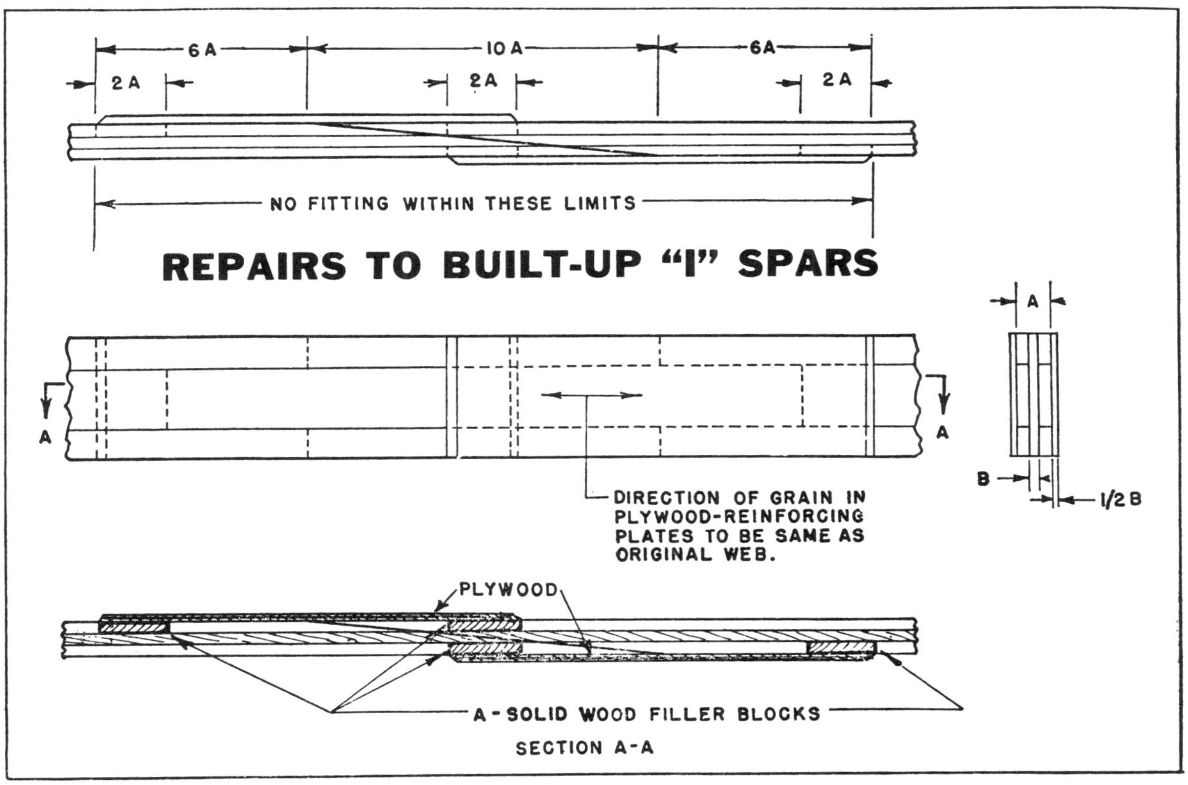

Your Forest Products Laboratory

By Bob Whittier, EAA 1235

EVERY SERIOUS student of aircraft design has at one time or another seen mention of the Forest Products Laboratory in reference texts. In view of the popularity and usefulness of wood in amateur-built airplanes, I recently made a visit to this Laboratory to see for myself what kind of a place it is, what kind of work it does, and what it is in a position to do to assist the present-day amateur airplane designer and constructor.

The laboratory is located in Madison, Wis. The capitol of that state, this city is an interesting and pleasant mixture of government, commerce and education, being also the site of the University of Wisconsin. Such a location is most favorable for a laboratory of this type, for there it is in a position to cooperate with the lumber industry on one hand and the academic world on the other.

The Forest Products Laboratory of the U. S. Department of Agriculture's Forest Service is located on a street that branches off from one of the city's main thoroughfares in a section that is heavily built up in neat and attractive apartment buildings, hospitals and various buildings of the university complex. As I drove down this side road I found myself leaving the built-up area rapidly behind me and not very far ahead I could see open fields and beyond them, wooded hills. My first sight of the Forest Products Laboratory took me very much by suprise; it proved to be a most impressive looking structure, built on high ground near the open fields and very well landscaped.

I walked up a flight of broad stairs leading to the main entrance and found myself in a large lobby which was liberally furnished with displays and exhibits of various things having to do with wood technology in its many phases. The receptionist was most courteous and in a few minute I was shaking hands with Mr. Edward W. Kuenzi, with whom I had an appointment and who was to guide me through the laboratory.

The present building was constructed in 1932, he explained as we began our tour, and its five floors contain 175,000 sq. ft. of floor area. About four hundred persons are employed, of which one-third comprise the technical staff responsible for research work. Over the years the laboratory has conducted a very large number of research projects having to do with every imagineable aspect of the utilization of wood.

On the lower floors of the laboratory building is located a wide variety of testing and experimental facilities. On the ground level in one wing of the building is the woodworking shop which is equipped with every kind of power tool used for cutting and shaping wood. This shop serves a dual purpose. It can cut wood to any dimensions needed for laboratory tests and can fabricate any kind of wooden structure for the same purpose. It can also perform research on woodworking tools and methods.

As an example of the latter, with the dwindling of the nation's forest resources they have studied methods of utilizing species of trees not hitherto commercially harvested. By developing suitable cutting tool tooth designs and operating speeds, they have made it possible to turn out acceptable lumber and finished products from species having peculiarities of grain and texture that had frustrated previous attempts to use them commercially.

In the same wing of the building I also saw the laboratory's mechanical testing facilities. My guide led me into one of a number of rooms each of which contained a vibration and fatigue testing machine. In the room we entered, a whirring vibration machine was endurance-testing a small strip of material which I was told was a boron type structural material they were testing for the space flight people.

This underlined the important point that facilities of the laboratory are available to any government or private agency having appropriate research or testing programs of importance enough to science and industry to justify assigning needed laboratory facilities to the work they wish to have performed. Nowadays the laboratory does a considerable amount of work for the aerospace program, particularly involving sandwich and honeycomb materials.

Not far from the vibration testing chambers was a battery of testing machines designed to bend and compress samples of wood and wooden structures. One in particular caught my attention. You have seen a screw-type arbor press in a machine shop? Well, imagine a sophisticated version of such a device, having a pair of screws as thick as the hydraulic column on a service station's grease lift and more than two stories high, and you have a general idea of the appearance and proportions of this mighty rig. It can apply a compressive force of one million pounds to a fabricated column, pile, post or anything else of that sort whose compressive strength had to be determined.

"What happens when a large specimen being tested in this machine breaks?" I asked.

"It makes a h--- of a noise!" grinned Mr. Kuenzi.

Next he led me to a long corridor down in the basement and showed me a series of atmospheric testing rooms in which temperature and humidity can be controlled to duplicate many climate conditions. We stepped very briefly into a room in which the temperature was well below zero and I saw samples of glued wood being tested to determine their durability under such conditions.

The room adjoining the cold one had its air maintained at a level of temperature and humidity typical of heated rooms in an average dwelling in wintertime. Separating the two rooms was a panel of insulating material which was being tested for effectiveness and vapor transmission properties under conditions that would exist in winter if it was used in wall construction.

A little farther along the corridor we came to another room in which the air was as hot and moist as in the most humid of tropical jungles. Looking through the droplet-flecked window I could see that the air inside was actually steamy! We entered and the dank smell of rotting wood assaulted my nostrils. Wood samples ranged on supporting racks were overgrown with rot fungus.

Other parts of the laboratory house the offices of research specialists and chemical laboratories where studies are conducted of such things as paints, preservatives and glues. There is a very complete library whose volumes on all aspects of wood science are available to interested groups or individuals. There's even an electron microscope used in studies of the basic structure of wood and the mechanics of impregnation with adhesives and preservatives.

Mr. Kuenzi led me across a parking lot to a newly completed building which looked to me as if it could house a large electric generating plant. Once inside, I saw that its vast interior was free of columns and supports, laminated wood arches being used in its construction. Work is currently underway, my guide explained, in installing and putting into operation a very modern pulp and papermaking facility. The Laboratory keeps right up to the minute with its work and nowadays is very active in the field of pulp and paper technology — paper being such an important material in today's world.

Back in Mr. Kuenzi's office after having seen the Laboratory's facilities and gained an overall familiarity with its various activities, I asked what they were doing these days in the field of aviation. Mr. Kuenzi said that, frankly, they are not doing a great deal of active research work. Factories which produce commercial and military aircraft have gone over more or less completely to metals and the only work currently being done in the Laboratory having any relationship to aircraft involves the determination of design criteria for construction of various materials.

To the best of his knowledge, he went on to say, nothing has been done in the last decade or two in the direction of making flyable experimental aircraft with sandwich material for the airframes. If someone came along with a technically sound and financially feasible proposal for some such project, he noted, they would probably be able to assist with it in one way or another for the sake of adding to the overall store of knowledge about how sandwich materials work out in practice.

Most of the Laboratory's work involving the utilization of wood in aircraft was done from ten to fifty years ago, he explained. Since new models in Sitka spruce trees and other species suitable for aircraft have not been introduced and we're therefore still using the same kind we started out with, information developed by the Laboratory in the past is still the best available and can be relied on by designers.

During the first World War when wood was the principal structural material for airplanes, the Laboratory did much work in evaluating various species of wood and the data thus developed found its way into aviation publications. Information on the properties of the more common acceptable substitutes for Sitka spuce, for example, is set forth in CAM-18, "Maintenance, Repair and Alteration of Airframes, Powerplants, Propellers and Appliances", copies of which are available from EAA Headquarters for $3.00 each.

I asked about two other publications, ANC-18 and ANC-19, published during World War two and still often sought by persons having a serious interest in wood as a material for light aircraft construction: Both were the result of an urgent wartime need to save metal for use in combat equipment and were meant as rather comprehensive and advanced guides for the engineering departments of aircraft factories undertaking to design and manufacture training and support aircraft of wood. Some of the information in these thick volumes is of interest and use to the average amateur aircraft builder, but much of it is on the theoretical and mathematical side to the extent that only a person with some formal engineering education can make good use of it.

Mr. Kuenzi informed me that the Laboratory's stock of these publications is now exhausted and it is beyond the Laboratory's scope to reprint them. No wide need now exists for them and there would have to be such in order to justify reprinting. However, a significant number of communications addressed to the Federal Aviation Administration by amateur aircraft designers and builders, emphasizing a need for them in the interests of safety, might produce results. In the meantime, copies should be available to serious students in many large engineering college libraries.

Despite this disappointment, there is much the Laboratory is willing and able to do to assist serious amateur aircraft builders. All the information developed by the Laboratory over the years is available to the public. Each year the Laboratory answers thousands of letters seeking general or specialized information, and problems are discussed with those who come to the Laboratory in person seeking help or information. General visitors are conducted through the Laboratory in groups at 2 P.M. each working day; most of these visitors, from 8,000 to 10,000 each year, come during the summer months.

In an average year about 3,600 consulting visitors representing all 50 states come to the Laboratory for assistance and about 400 more come from other countries. Normally, these people are trained persons representing some enterprise or agency and they come in person after engaging in correspondence with appropriate specialists at the Laboratory. When such a person with a specific and serious interest visits the Laboratory, every effort is made to give him the assistance he needs. Where the problem presented is of such scope and difficulty as to warrant a cooperative research project, the work will be undertaken if it is consistent with the Laboratory's public objectives. An advance agreement is made as to methods to be followed and the payment of costs.

The Laboratory's guiding purpose in such studies is to secure facts that will promote the best use of wood. A pamphlet explaining this cooperative service more fully can be obtained on request to the Laboratory.

The Laboratory at this time has three publications lists which are of particular interest to EAA members and which can serve as a useful springboard into making good use of the Laboratory and its services. They are, "Publications of Interest to Architects, Builders, Engineers and Retail Lumbermen", "Publications on Glue, Glued Products and Veneer", and "Publications on Sandwich Construction", each of which is free. Requests for publications, lists and information should be addressed to the Director, Forest Products Laboratory, Madison, Wisconsin 53705.

Scarfing Sander

By Wesley H. Buckman, EAA 15371

Here's a little gadget I came up with while searching for an easy way to scarf plywood for my project. It's simple, inexpensive and, above all, foolproof! All the other methods I've seen involve feeding the plywood into a saw or some other piece of equipment. This way, the plywood remains stationary.

All edges to be scarfed must be well supported to insure an even cut. This can be easily done with an extra sheet of commercial plywood (¾ in.) or by using the edge of the workbench. Clamp or otherwise secure the wood being scarfed so that it cannot move. Practice first on some scrap to get the hang of the thing.

I have been able to get scarf joints nearly three inches wide on ⅛ in. plywood with no trouble at all, which I'm sure will make the inspector quite happy.

All that is needed to make this thing is a little scrap plywood, an electric drill (you're building an airplane without one??) and a sanding drum easily purchased at any Sears store. The cost is peanuts, but the time and effort saved are worth a fortune.

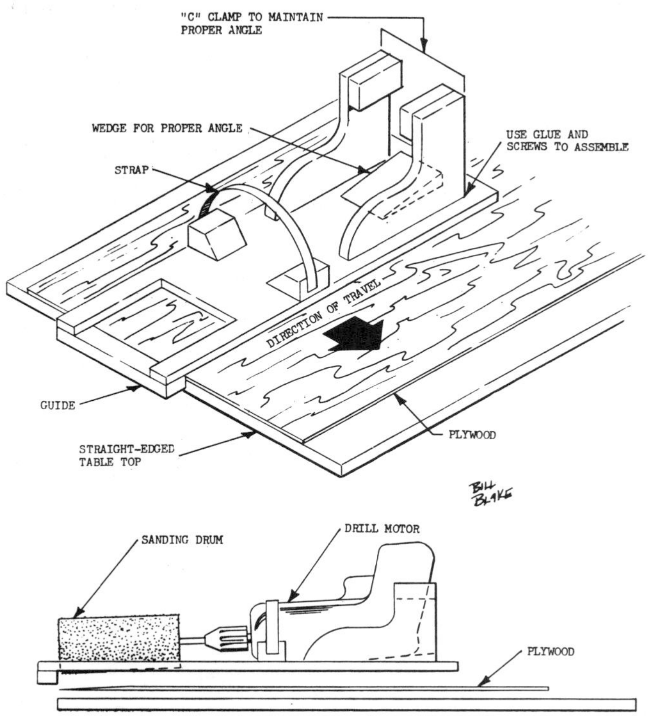

Drawing shows construction details of a[illegible] all wood French Formula 1 racer — th[illegible] MP-205 "Busard". Empty weight is 52[illegible] lbs. With a 90 hp Continental the aircra[illegible] will cruise at 180 mph.

MP-205 Busard

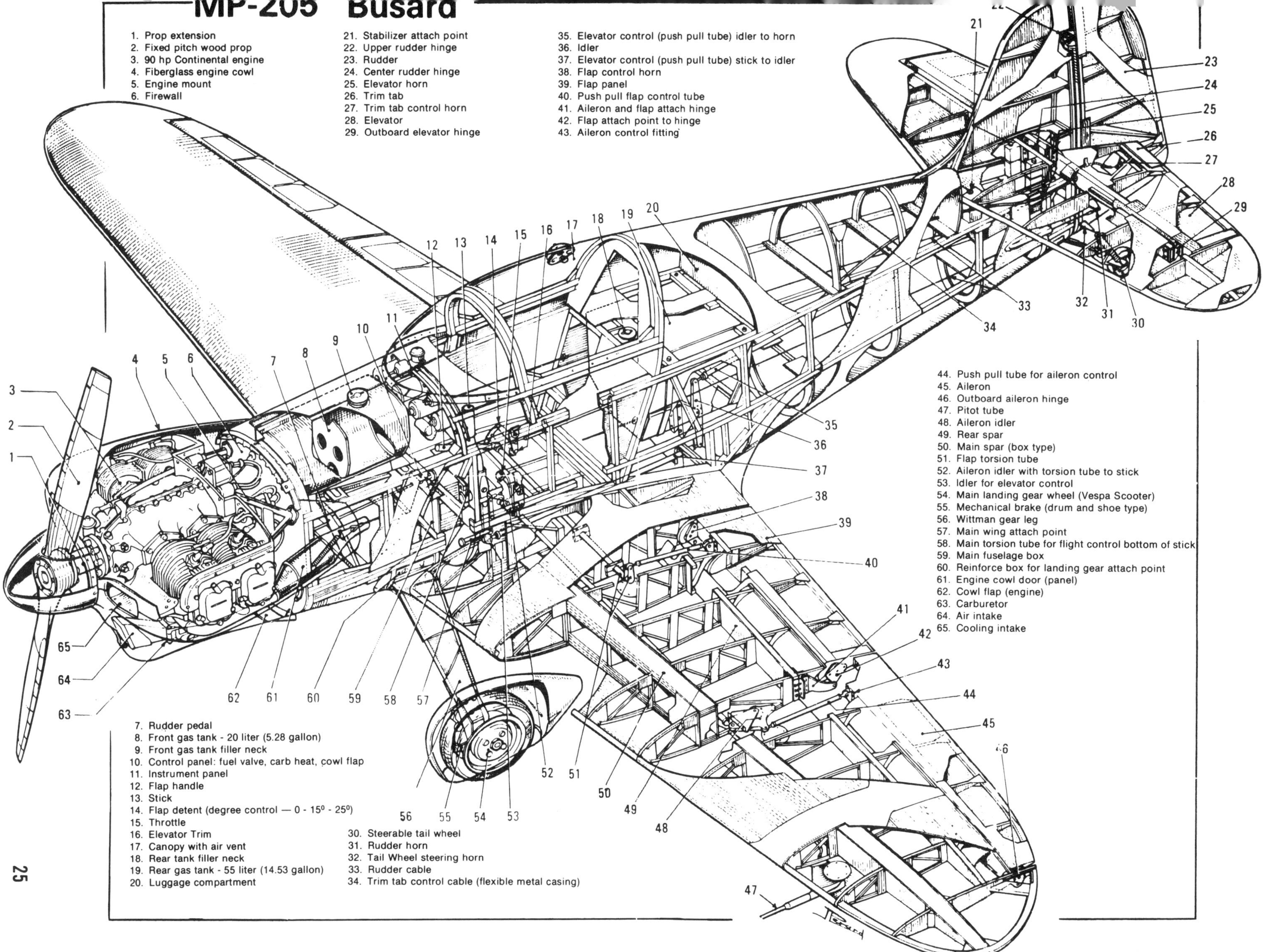

1. Prop extension
2. Fixed pitch wood prop
3. 90 hp Continental engine
4. Fiberglass engine cowl
5. Engine mount
6. Firewall
7. Rudder pedal
8. Front gas tank - 20 liter (5.28 gallon)
9. Front gas tank filler neck
10. Control panel: fuel valve, carb heat, cowl flap
11. Instrument panel
12. Flap handle
13. Stick
14. Flap detent (degree control — 0 - 15⁰ - 25⁰)
15. Throttle
16. Elevator Trim
17. Canopy with air vent
18. Rear tank filler neck
19. Rear gas tank - 55 liter (14.53 gallon)
20. Luggage compartment
21. Stabilizer attach point
22. Upper rudder hinge
23. Rudder
24. Center rudder hinge
25. Elevator horn
26. Trim tab
27. Trim tab control horn
28. Elevator
29. Outboard elevator hinge
30. Steerable tail wheel
31. Rudder horn
32. Tail Wheel steering horn
33. Rudder cable
34. Trim tab control cable (flexible metal casing)
35. Elevator control (push pull tube) idler to horn
36. Idler
37. Elevator control (push pull tube) stick to idler
38. Flap control horn
39. Flap panel
40. Push pull flap control tube
41. Aileron and flap attach hinge
42. Flap attach point to hinge
43. Aileron control fitting
44. Push pull tube for aileron control
45. Aileron
46. Outboard aileron hinge
47. Pitot tube
48. Aileron idler
49. Rear spar
50. Main spar (box type)
51. Flap torsion tube
52. Aileron idler with torsion tube to stick
53. Idler for elevator control
54. Main landing gear wheel (Vespa Scooter)
55. Mechanical brake (drum and shoe type)
56. Wittman gear leg
57. Main wing attach point
58. Main torsion tube for flight control bottom of stick
59. Main fuselage box
60. Reinforce box for landing gear attach point
61. Engine cowl door (panel)
62. Cowl flap (engine)
63. Carburetor
64. Air intake
65. Cooling intake

Wood Testing

. . . the Easy and Reliable Way

by S. O. Jenko

The strength of metallic materials used in aircraft construction is well known and, with other physical properties, it is defined by various specifications. The manufacturer must hold all physical properties within specified values, thus assuring uniformity of materials.

Wood is a natural product of which its properties are not only influenced by the growth of the tree but also by its processing, storage, and final moisture content. These factors also influence the strength of wood. In many cases the designer, as well as the builder, is forced to use just one "certified" kind of wood while others may not be considered because they are not certified.

While a few excellent articles published in past issues of *SPORT AVIATION* about the advantages of using wood species other than Sitka spruce demonstrated valid reasons, the problem of a simple and reliable test remained unsolved. Thus, the testing of the wood to determine its strength became of prime importance.

The structural testing of materials is a part of engineering mechanics. It required considerable theoretical understanding, practical experience, and usually expensive testing machines or set-ups. These things are usually beyond the capabilities of most homebuilders whose prime interests are to build and fly their own aircraft rather than to engage in elaborate and expensive testing. On the other hand, nothing is worse and more detrimental than an improperly conducted or a false test.

While attempts have been made in the past to design a simple testing device for testing aircraft materials such as wood, plywood, welds, and fabric, one early attempt deserves to be mentioned.

During the late 1930's, when the homebuilding of sailplanes in Europe became very popular, the German Research Institute for Soaring developed a universal testing "machine" built of wood in a typical aircraft-construction fashion. The plans were published (1940) in at least one book which was and still remains the most outstanding and authoritative work on wood aircraft-and-hardware building methods. However, this universal testing machine has two drawbacks; it is rather large (about two-feet wide, four-feet high, and eight-feet long), and it requires specially made test specimens which, in turn, are time-consuming to make.

Since then, other attempts have been made, some of them published in *SPORT AVIATION*, to solve the old problem of finding the strength of wood to be used in aircraft construction.

Several years ago the need for a simple, reliable wood-testing device became quite apparent when designing our high-performance IBAS amphibious aircraft. Since this aircraft was designed for experienced homebuilders and, in order to assure the required structural integrity, the plans clearly specify the minimal ultimate strength (psi.) of wood used as well as the corresponding wood density (pcf.). In order to avoid an avalanche of mail inquiries from amphibian enthusiasts about our IBAS, we would like to mention that the plans are not as yet available.

In addition to the above-mentioned requirements, two very important features were stipulated:

(A) Repeatability of testing procedure, i.e., once the testing device is aligned it should constantly provide the same loading conditions thus assuring comparative evaluation of test results;

(B) Usage of simple test specimens (comparable in size to most aircraft components) which can be made in a very short time.

While the design of a wooden aircraft is primarily based on the comparative strength of wood (since the compressive strength is considerably less than the tensile strength and it also governs the building) it becomes highly desirable to establish this important value.

Unfortunately, a direct approach would not only necessitate high loads (dead weight) to be used in the testing device but it would also require the end-grain surfaces of the test specimen to be exactly parallel, another nuisance and difficulty for the homebuilder. However, since

FIGURE 1

there is a relationship between the bending strength and the compressive strength of wood, one has two basic choices of design:

(A) Cantilever beam (see Fig. 1).

(B) Beam on two supports (see Fig. 2).

Due to a number of practical and theoretical considerations, the cantilever beam cannot compete with the beam on two supports. This arrangement was finally selected for out testing device. While we are experienced in design of complex testing devices, we soon found out that at least two requirements—simplicity with low cost, and ease of construction—presented considerable problems.

These and other design problems were successfully solved. Thus, we believe that a significant contribution to the safety of wooden homebuilt aircraft has been made. Any homebuilder interested in knowing the strength of the wood used in construction of his aircraft can find out in a short time and without any specialized technical knowledge

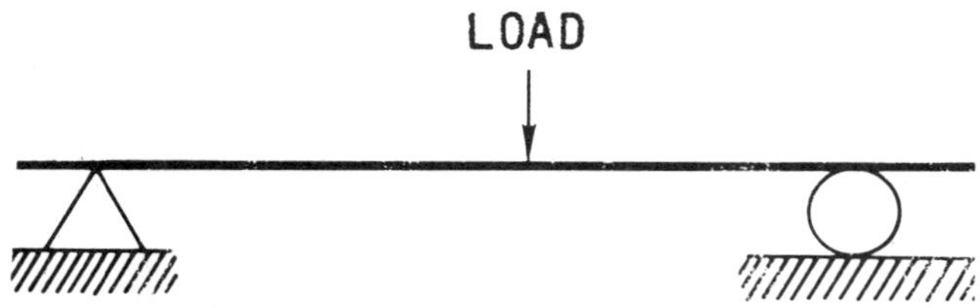

FIGURE 2

or calculation if the wood meets the specifications.

On the other hand, the cost of materials to build our testing device should be less than $10.00, an insignificant investment considering the benefits of having the peace of mind and sound structural material in the aircraft.

Fig. 3 shows the testing device. Considering that the test specimen is a simple strip of wood 3/4x3/4x14-inches long, the small size of the testing device becomes apparent. It should be clamped at the edge of a sturdy table or workbench as shown in Fig. 3. The specimen is inserted, the predetermined load (dead weight) is attached at the end of the loading bar as indicated, and then the handle is rotated thus gradually loading the specimen in the center.

The widely held belief that the specimen must break under the applied load to establish its ultimate strength is quite erroneous. The following explanation is in order:

When the load is applied at the center of the test specimen, it is loaded in bending, i.e., the topmost fibers experience the maximal compressive stresses, the bottom-surface fibers will show the maximal tension stresses. Thus, when the load is increased to a certain value the topmost wood fibers will collapse and the capability of carrying the load is gone. This collapse of topmost fibers will rapidly advance toward the neutral line of the test specimen cross-section (longitudinal centerline) if the load is slightly increased, resulting in a complete break-up of the test specimen.

While this explanation may sound complex, it takes only a little diligence and a few trial tests to learn how to detect the collapse of the topmost wood fibers and thus determine the corresponding load. When this ability is acquired, the actual testing becomes a routine procedure.

However, the detection of the topmost fiber collapse often requires very close visual observation of this area. Of course, the test specimen once used cannot be reused when the collapse of the topmost fibers develop.

Because the moisture content of wood is one of the factors governing the strength of the specimen taken out of a board, it must have a certain weight. It varies with wood species. Thus, in addition to our testing device, a good quality letter scale is essential. It should be calibrated in order to ascertain its accuracy. Also, a table is needed showing loads to be applied at the end of the loading bar which produce failure in the topmost fibers of the test specimen and also its weight since both vary with the wood species.

The cross-sectional dimensions of the specimen must be as close as possible to those specified since any deviation will cause erroneous test results. The wood surface under the applied load must be very smooth in order to facilitate the detection of the topmost fiber collapse.

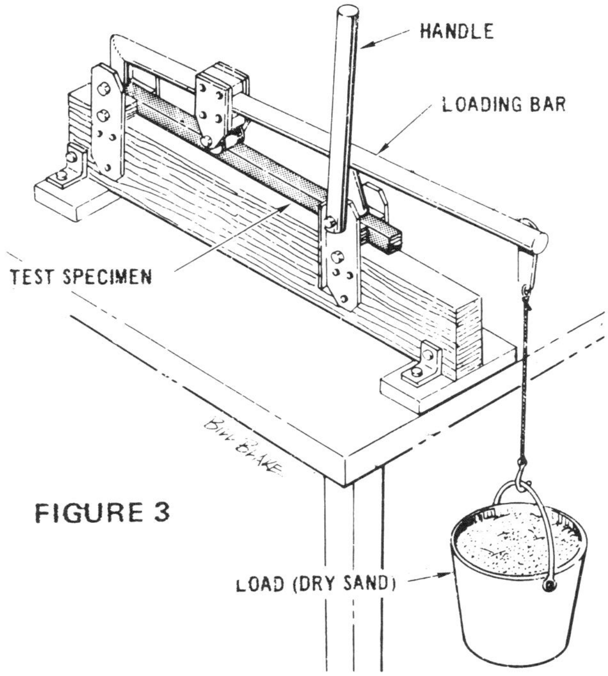

FIGURE 3

Getting Smooth Cuts

Neatness and accuracy being as important as it is in aircraft work, there is much interest in proper tools and methods for getting clean cuts in wood. Most of us have used ordinary table saws with rip and combination blades and have been unhappy with the rough cuts which result. Yet, it is quite possible to get cuts with a table saw that are so smooth that a few light passes with sandpaper afterward will eliminate tooth marks nearly 100 percent.

Low-price blades are often of thin metal and blades made for general ripping work are also quite thin so that their kerfs will be as narrow as possible with subsequent savings of wood. The trouble with any thin blade is that it is apt to "chatter" at high speed and thus throw tooth marks into the wood. It can also produce wavy cuts by reason of a tendency to follow the grain. Ripsaws almost always have some set to their teeth, to make the kerf wider than the blade and minimize binding, warping and crooked cutting.

The correct saw to use for aircraft work is a "cabinet maker's blade". They come in different designs; some are hollow-ground with thick teeth, thin disc, and thick hub. Some have thick hubs with a step-down out near the teeth to keep kerf reasonably narrow. The teeth have no set to them and usually are quite thick as compared to the lighter general purpose blades. Some have very small teeth but others have fairly large ones, while retaining the essential feature of no set and a thick, stiff disc.

When cutting long, thin strips it is desirable to make some kind of guide, perhaps of wood with spring "fingers", which will hold the wood snugly against both the table and fence and keep it from bending. When wood bends or chatters, tooth marks are put into it. Feed wood in at as nearly uniform a rate as possible because halts and jerks also make tooth marks.

Many firms make good cabinet maker's blades but to get EAA members started on the right track it can be mentioned here that Sears, Roebuck and Co. has a "Thin Rim—Satin Cut Combination Blade", Cat. No. 9-3254, which upon trial has produced wonderfully smooth cuts in spruce, fir, pine and mahogany.

Owners of cabinetmaker's blades should not use them to cut plywood. The glue lines in plywood are surprisingly hard and brittle and can dull a fine saw rather quickly. Special blades for cutting plywood are also available, which feature very small teeth to keep edge splintering to a minimum.

Quantity Production of Quality Wooden Parts

by Chuck Larsen, Designee Director

During the construction or restoration of an airplane numbers of parts made of wood must be accurately fabricated. An example of this is making noseblocks for a wooden rib. This kind of part can be accurately duplicated in quantity by making a template and finish shaping it using a flush laminate trimmer as a bearing mounted on the end that runs against the guiding edge of the template. The photos illustrate this tool in use.

The template is constructed by making a full scale drawing of the part on a sheet of ¼" aluminum or hard plywood. The template should be about 2" longer than the finished part to provide for the clamps that hold the template to the material while the part is being trimmed on the router. The excess material is trimmed from the part to finish it. The template MUST be carefully cut and finished — the final product can be only as accurate as the pattern from which it is made. The time spent in making the template should be regained in the third or fourth part made from it. If you are making a biplane with 40 ribs or the template can be used by more than one person both the quality of the parts and the time necessary to produce them will be improved. This system can be used effectively for making wooden parts requiring precise duplication in any quantity.

When the template has been fabricated, the laminate trimmer secured and the router fastened to a workbench or router table the following steps should result in the "Quantity Production of Quality Wooden Parts".

1. Using the template, outline the part on the material from which it is to be made.
2. The part is then "rough cut" with a saber or band saw approximately ⅛" from the scribed line.
3. The template is then firmly clamped over the scribed part with small "C" clamps.
4. The depth adjustment of the router should be set so the bearing on the laminate trimming bit runs on the template while the cutting edges meet the entire surface to be trimmed.
5. The part can then be trimmed to its final size and shape with the router.
6. The excess material used for clamping is then removed to complete the part.

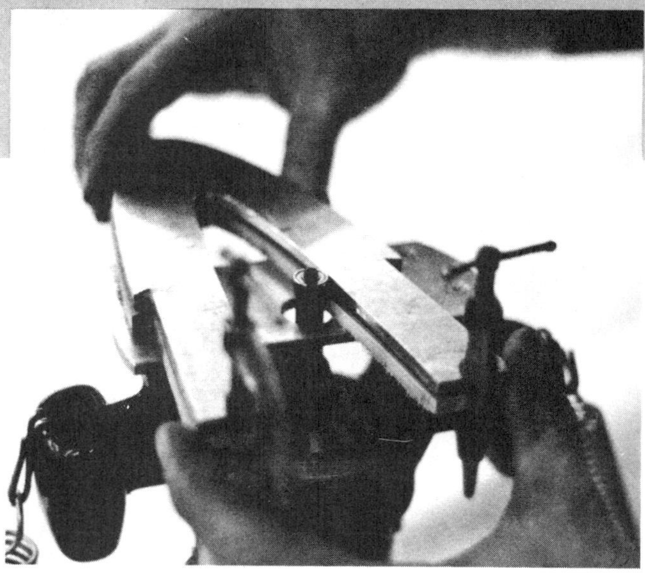

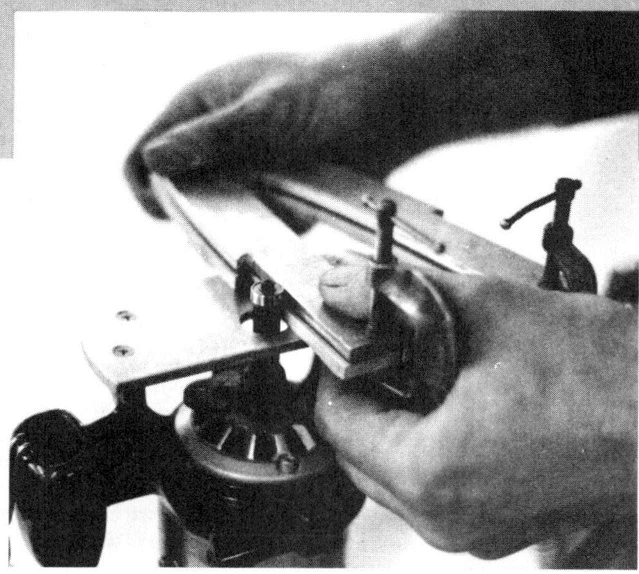

How to Form Plywood for a Leading Edge

By Walter W. Redfern, EAA 143

Ordinary rubber inner tube, cut into strips 1 or 1½ in. wide, can be strapped around the leading edge of your plywood covered wing as shown in the accompanying sketch.

As the rubber bands tighten up, the plywood conforms to the shape and will be held there to dry. Where more pressure is required, several of these rubber bands can be applied. The radius of the leading edge will determine the soaking time for the plywood.

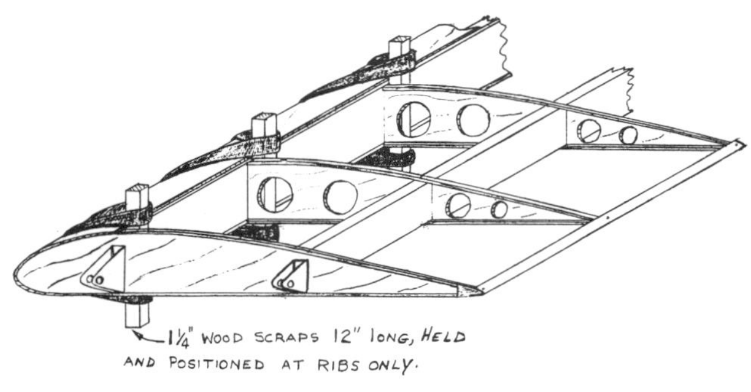

Variable Miter Head

By Leon "Jerry" Jones (EAA 100301)

I am building the beautiful and fast Barracuda.

As I started construction, I became acutely aware of the shortcomings of my miter head which mounts on my table saw. Its support starts out a couple of inches from the saw blade and when you rotate it to various angles, the supporting edge moves progressively further from the saw. This means that every piece of cut wood has to be a foot long — or has to be clamped to the cross head in order to support what is being done at the saw blade.

Since "necessity is the mother of invention", I have designed the variable miter head shown in the accompanying photographs (Figures 1, 2 and 3). It offers a number of advantages:

FIGURE 1

FIGURE 2

(1) An adjustable stop so that repetitive cuts can be made accurately.

(2) Since the pivot is mounted within ½" of the saw blade, rotating the miter head to practically any angle does not move the piece being cut away from the saw blade. It is very easy (providing you don't have nervous fingers) to cut a 60⁰ angle on one inch lengths of material.

FIGURE 3

(3) Angles can be transferred from an adjustable protractor to the miter head with a high degree of accuracy. This means you can measure an angle right on your airframe and transfer it to the miter head. The adjustable stop can be moved in or out in very small increments to bring your finished part into a perfect fit.

A good example of the usefulness of my variable miter head is the way I built the wing ribs for my Barracuda. I made up the first rib, taking care to get every piece accurately fitted. I then dismantled it, transferred the various angles and lengths to the table saw and cut all 960 individual rib pieces for all 18 ribs . . . in one evening. They all fitted perfectly when I made up the remaining ribs.

Treat That Wood

By Richard B. Anderson (EAA 5432)

WOOD AIRCRAFT STRUCTURES are prone to wood rotting fungi in structures that collect water from rain or splashing, and in areas of continued high humidities. Finishes will seal wood and retard moisture absorption but do not prevent it. Wood will absorb moisture which includes humidity. Once moisture is absorbed, the wood finish will also retard drying allowing high moisture conditions to exist for a period of time.

Fungi that rot wood require moisture and warm temperatures for best growth. These conditions will be found in wood aircraft structures in most parts of the world and should be guarded against for extended aircraft life. Good aircraft design and preservative treatment along with good finishes will offer an insurance policy against wood rotting fungi.

The Forest Service, U.S. Department of Agriculture, has published "Technical Bulletin No. 1334, Preserving Wood by Brush, Dip, and Short-Soak Methods" for sale by the Superintendent of Documents at 25c each. This bulletin has application for aircraft structures although tests were carried out on boxes, porch steps, and columns.

Pentachlorophenal 5% solution with a lightweight proven solvent of the aliphatic mineral spirits type is recommended. Brush and 3-minute or shorter dip treatments will afford good protection in aircraft designed to avoid the worst decay hazards. The bulletin recommends temperatures of the treating solution to be as low or lower than the wood to be treated for better penetration and absorption.

Pentachlorophenal is available at most hardware and farm stores. The solution should be applied liberally to keep wood wetted. The wood should then dry for 48 hours or more before painting. The best preservative will be only as effective as the amount penetrating the wood in relation to the severity of exposure. Brush, dip, and short soaks give only limited pentrations; therefore, they offer good protection only under low to moderate hazards. Aircraft construction would normally be a low to moderate hazard and preservative treatment followed by painting will give good protection that cannot be acquired by painting alone.

Do not forget holes and areas under metal fittings.

Some commercial preparations contain a waterproofing compound. A test structure should be treated and then finished with type of paint or varnish to be used to test adherence and compatability.

Duplicate Sawing

By John W. Grega (EAA 3808)

I've used this method in wooden aircraft production during and after World War II and always found it very useful and time-saving, along with the minimum amount of tooling required.

The guide is made of 1" x 3" pine, cut to a length about two-thirds the length of the rip fence on your tablesaw, the two pieces of 1" x 3" are nailed right angle to each other as in Figure 5. The guide is clamped to the rip fence using two 3" "C" clamps, with the guide raised above the saw table the thickness of the work material, plus 1/16" clearance for the work piece to slide smoothly under the guide.

If a number of pieces are to be duplicated, I prefer making the patterns out of ¼" Masonite, smoothed to contour on a sanding disc. Accuracy counts in this area, as any discrepancy in the pattern will be duplicated on the work piece.

Fastening the pattern to the work piece can be done in a number of ways, I use a ½" long, thin wood screw, round head or flat head, screwed through the masonite pattern, then using a mill file, sharpen the end of the protruding screw to a sharp point. In this manner, the pattern can be pushed, by hand, into the work piece. Do not worry about the marks in the work piece, made from the spurs, as these tool marks are acceptable to FAA inspection as long as they are small in size and not too deep as to split the work piece. Other methods are to use old time phonograph needles or aircraft type nails to make the spurs.

With the guide clamped in place on the rip fence, adjust the rip fence so that the guide edge is flush with the saw blade, as in Figure 4, then crank the saw blade upwards, with the motor running, and slowly feed the saw blade into the guide about halfway or ⅜". A good test is to run a scrap piece of material against the guide with the motor running. The saw blade should just barely make a mark in the scrap piece, if it cuts into the scrap piece, adjust the rip fence towards the saw blade until the cutting sound stops, this way our saw blade will not cut into the pattern.

We are now ready to produce any number of duplicate curved pieces, such as ribs, nose rib pieces, formers, etc. With the pattern attached to the work piece, a number of straight cuts are made against the guide to get rid of all excess material, then a smoothing pass is made following the pattern. No sanding is required if you use a hollow ground saw blade. I like the Sears Hollow Ground Combination Blade, Catalog No. 9M3239. Incidentally, as you are cutting and making a number of passes, do not neglect to stop the saw now and then, and clean away the pieces that are accumulating under the guide, otherwise, they will jam up and bind against the saw blade.

Styrofoam ribs are a cinch to make, using the above setup, I cut the nose, center, and trailing edge ribs for the PDQ aircraft in one hour. The time saving is substantial.

Another use for the guide setup, and a money saver, too, is to purchase Spruce or Fir in the S2S, or Surfaced Two Sides, for wing spars. In this condition, only the two faces are machined to thickness, the width remains as received from the mills and the edges are rough cut, therefore none too straight. All we have to do is straighten one edge, then it can be ripped to finish size. Doing this operation on a joiner is both time consuming, and a waste of expensive material. One saw cut is all that is needed to give us a straight edge suitable for aircraft spars.

To accomplish this, using aircraft nails, fasten a piece of Pine about ¼" thick and 1" wide, one nail at each end, to the one rough edge of the spar material, Figure 1, then sighting down the one edge that will contact our saw guide, straighten the piece of Pine by hand and nail it in a few places until it sights straight, then place it on the saw table, start up the motor, and with the Pine piece up against the guide, give it one pass and your material has one straight edge. Remove the Pine piece, remove the guide, adjust the fence for the finished spar height, and rip the spar to size. The ripping or off fall, can be utilized, for cap strip, corner blocks, etc. and at $1.50 a board foot, every piece can be used, saw dust or shavings can only wind up in the rubbish barrel.

CUTTING TAPERS: Lay out the taper, using a lead pencil, on the spar material, tack the ¼" x 1" Pine strip, following the pencil line, and cut as previously described. If you need a pair of tapered spars, like a stabilizer and elevator spar, tack the first spar you made, on top of the second spar blank and cut away. This results in two identical tapered spars.

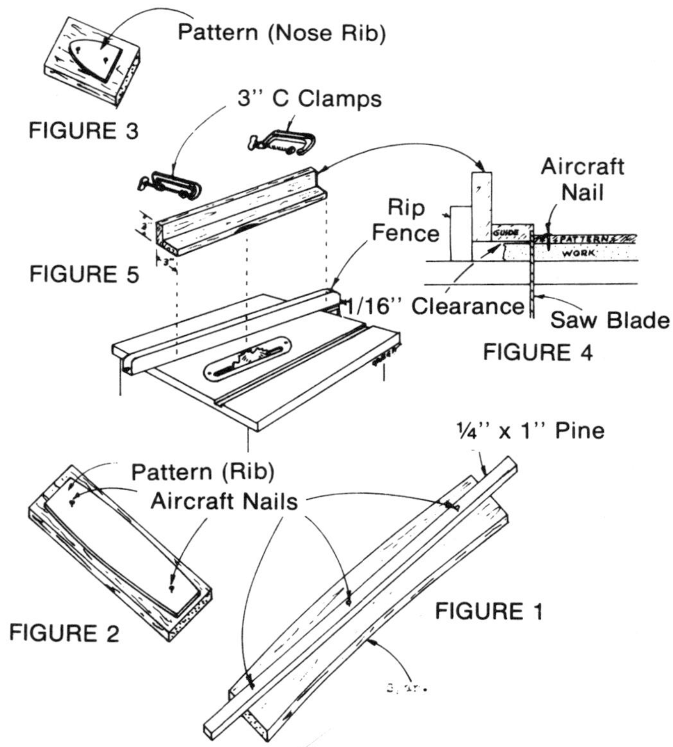

Characteristics of Aircraft Type Glues

RESORCINOL-FORMALDEHYDE GLUES.

Resorcinol-resin glues are furnished in water-alcohol mixtures that are stable throughout more than a year of storage at room temperature. When mixed with a hardener and ready for use, the solids content of these glues ranges between 60 and 70 per cent. The resorcinol resins are classified as room-temperature-setting and require a pressure period of 5 to 7 hours at 75° F., although full joint strength is not usually developed in this period. Resorcinol resins should not be used at lower temperatures, and their rate of cure is accelerated at higher temperatures. The working life of resorcinol resins ranges from 2 to 5 hours at 75° F. and is considerably shortened at higher temperatures. The disadvantage of short working life in warm rooms can be overcome by surrounding the glue container with a water bath in which cool water is circulated to maintain the temperature of the glue mixture at from 70° to 75° F. or by keeping the mixed glue under refrigeration until required for use. Resorcinol glues are similar to the intermediate-temperature phenols in their ability to produce strong joints in wood over a wide range of moisture content. Assembly periods for resorcinol glues vary somewhat with different glues, but they are, in general, shorter than for intermediate-temperature phenols, and maximum closed assembly periods range from 50 to 120 minutes at 75° F.

MELAMINE-FORMALDEHYDE GLUES.

Melamine glues are available in hot-press and intermediate-temperature types and in film and powder forms. With some exceptions, their use characteristics are similar to those of the phenol-formaldehyde glues of the same type and form. Melamine resins are stable over long periods and commonly have a storage life of at least 1 year at room temperature. The temperatures required for curing the hot-press melamines are usually somewhat lower than for hot-press phenols.

UREA-FORMALDEHYDE GLUES

Urea-resin glues are marketed either as dried powders or as water suspensions ordinarily containing 60 to 70 percent of solids. Most of those marketed as liquids may be used as received or after the addition of small amounts of catalyst, which increases the rate of setting. The dry-powder types are mixed with water to produce suspensions having concentrations approximately the same as those marketed as liquids. In general, urea-resin glues in powder form have a longer storage life than those in liquid form. Some urea-resin glues are formulated for use at room temperatures and others for hot-pressing operations.

Urea-resin glues appear to be more critical than resorcinol resins with respect to the minimum moisture content of the wood at which strong glue joints can be made, although the most desirable moisture content will ordinarily be governed by considerations other than its effect on the quality of the joint. When hot pressing with urea-resin glues at high temperatures, the same limitations on moisture contents must be observed, in order to avoid blistering, as with phenol-resin glues. Assembly periods are not critical with hot-press urea-resin glues and may be varied over comparatively wide limits, in some instances for more than 24 hours. Urea-resin glues formulated for hot pressing set at temperatures somewhat below those required for phenolic-resin glues.

The rate of setting cold-setting urea resins at room temperatures is much slower than that in hot presses and a minimum pressure period of 4 hours at 70° F. is required. At higher temperatures shorter pressure periods may be used. Curing continues at room temperature following the release of pressure. The use of cold-setting urea-resin glues when the temperature of the wood or the gluing room is below 70° F. is prohibited by Specification AN–G–20. The working life of cold-setting urea-resin glues may vary from 2 to 8 hours at 70° F. but is considerably shortened at higher temperatures. In hot weather it may be desirable to cool the mixture by means of a water bath as described for resorcinol-resin glues. Closed assembly periods for cold-setting urea resins are more limited than for room temperature-setting resorcinols. Closed assembly periods should not exceed 20 minutes at ordinary room temperatures. Open assembly periods are generally limited on the basis of the relationship that 1 minute of open assembly is equivalent to 2 minutes of closed assembly.

"FORTIFIED" UREA-RESIN GLUES.

Other urea-resin glues have been developed for special purposes. Some of these glues, for example, have been developed primarily to improve the resistance of urea-resin glue joints to boiling water. Other formulas have been developed primarily for the gluing of curved plywood by the bagmolding technique, and these glues are characterized by the property of permitting very long assembly periods, although some will set quickly at temperatures at or near 212° F. Such glues are sometimes marketed under the term "fortified urea-resin glues," indicating that through the addition of resorcinol or other fortifying ingredients their resistance has been increased in some particular over that ordinarily associated with urea-resin glues.

CASEIN GLUES.

The use characteristics of casein glues have been reported in several publications. Their use over many years has generally resulted in good shop practices. In working life assembly time, pressure, and rate of setting at ordinary room temperatures they are similar to cold-setting urea resins and in general are subject to the same limitations and adapted to the same gluing operations. They do, however, set at temperatures lower than those at which the cold-setting urea resins can be used. The addition of preservatives to casein glues may affect somewhat the consistency or viscosity of the mixture and the working life, but, in general, the use characteristics of casein glues containing preservatives are similar to those of casein glues without preservatives.

Laminated Spars

By Bob Gilbert

THE AMOUNT OF PERFECT lumber of any species in the larger sizes used in aircraft is limited, but greatly increased quantities of suitable material may be obtained by the judicious utilization of smaller sizes through the technique of spliced and laminated construction.

A lamination may consist of a single piece made of two or more pieces edge-glued to form the required width or depth. Edge-glue lines in adjacent laminations should be staggered not less than the thickness of the thicker lamination.

WHY EDGE GRAIN?

An edge-grained board is one in which the annual rings make an angle of 45 degrees or more with the wider surfaces. Requirements for the use of flat-grained or edge-grained material are based on consideration of the stability of dimension and shape, and not because of any difference in strength. Edge-grained lumber shrinks and swells less in width than does flat-grained lumber. Consequently, changes in the vertical dimension of a spar and changes in moisture content, during manufacture as well as in service, are minimized by making its vertical face edge-grained. In general, the use of flat-grained and edge-grained laminations in the same assembly should be avoided.

GRAIN SLOPE

In general, the slope of grain in any part with respect to the longitudinal axis of the part should not be steeper than one in 15. The slope of grain within the middle half of the depth of solid or laminated spars may be permitted to be as steep as one in ten. Local grain-slope deviations in excess of one in 15 may be permitted in spar flanges and horizontally or vertically laminated flanges. In this case we are speaking of flanges for box or I-beam spars.

RULES FOR SCARF JOINTS

The following requirements apply to all scarf joints in solid or laminated aircraft parts:

1. The slope of the scarf should not be steeper than one in 15.
2. All scarf cuts must be made in the general direction of the grain slope.
3. In laminated members the longtitudinal distance between the nearest tips of scarfs in adjacent laminations should not be less than ten times the thickness of the thicker lamination.
4. The number of scarf joints should be limited as much as possible.
5. The location should be limited to the particular portions of a member where margins of safety are high and stress concentrations are not serious.

MOISTURE CONTENT

Each piece of lumber at the time of fabrication should have been dried to an average moisture content of not less than eight or greater than 12 percent.

GROWTH RINGS

The number of annual rings in any one inch measured in a radial direction on either end section of a lamination or part should not be less than eight for Douglas fir, six for white pine, six for Sitka spruce, and six for white spruce. If the radial dimension of the piece is less than one inch, there should be a proportionate number of rings.

DENSITY

The specific gravity should be at least the following: .45 for Douglas fir, .34 for white pine, .36 for Sitka spruce, and .36 for white spruce.

REQUIREMENTS FOR HORIZONTALLY LAMINATED SPARS

A horizontally laminated spar is a spar in which the cross section is made up of two or more laminations glued together and in which the principal glue planes are horizontal.

1. Laminations should be edge-grained on those faces which will be vertical in the finished spar.
2. Within either outer quarter of the spar depth, the size of the knot in a lamination where the vertical dimension is greater than one-eighth, the spar depth should not exceed one-sixteenth the width of the spar. The sum of the sizes within a length of the lamination equal to 5W (W = spar width) should not exceed 1/8W, and the sum within a length equal to W should not exceed 1/16W. Within either outer quarter of the spar depth, the size of a knot in a lamination where the vertical dimension is one-eighth the spar depth or less, should not exceed 1/10W, the sum of the sizes of all knots within a length of the lamination equal to 5W should not exceed 1/5W, and the sum in a length equal to W should not exceed 1/10W. Within the middle half of the spar depth, the diameter of a knot in any lamination should not equal 1/2W and the sum of all knots in a length of the lamination equal to W should not exceed 1/2W.
3. A pitch or bark pocket in any lamination should not be deeper than 1/8W or one-half the vertical dimension of the lamination, whichever is lesser, not wider than 1/4 in. or 1/8W, whichever is the lesser, and not longer than two inches with the further requirement that a pocket on a face of the spar should not be longer than four times its distance from a corner of the spar. The distance, measured in any direction, between two pockets on the same face of the spar should not be less than six times the length of the shorter pocket, and for pockets in the same growth layer this distance should not be less than six times the length of the longer pocket.

Substitute Woods and Simple Testing Rig

by Ron Scott

The cost of aircraft construction materials has been on the rise increasingly during the past few years. The manufactured materials, steel and aluminum, have risen in cost dramatically. Sitka spruce on the other hand has increased at a much slower rate than the above mentioned, but in most areas is difficult to obtain locally. If budget is a factor, we find that there are several domestic woods that can be substituted for Sitka spruce in varying degrees. NACA Report 344 (formerly ANC-18), Design of Wood Aircraft Structures, is a very good reference relative to wood strength in aircraft construction application. This manual is available from National Technical Information Service, 5285 Port Royal Road, Springfield, VA 22151. Douglas Fir is one of these locally obtainable domestic woods and prompts this fine article and illustration by Charles J. Barrows, EAA 39385, 46 Summer St., Kennebunk, ME 04043.

MORE ON THE USE OF DOUGLAS FIR AND A RIG TO SAMPLE BEND TEST

I have been building a Jodel F-12 and have used Douglas Fir as obtained from selected pieces of flooring stock which comes planed ¾" thick and usually 3½" wide, although sometimes greater widths are obtainable. The pros and cons of using non-certified wood have been covered in past articles in *SPORT AVIATION* and will not be covered here.

I purchased a fair quantity of this wood several years ago and, testing it as per the article presented by Don Cookman on page 15, December 1961 *SPORT AVIATION*, I found failures to be extremely rare. I recently purchased more stock and have been dismayed to find the failure rate approaching or passing 50%. Investigation elicited the information that B-and-Better, the best grade of wood to be found in lumber yard stock, is rarely to be found in flooring any more. Also, the Cookman method has the weakness that the full length of stock cannot be tested unless an almost infinite number of tests are made along the length of the sample.

Faced with this problem, I have made a testing rig which will test the full length (except that not covered by the moment arm) of a sample ¾" x ¾" X-section cut from the full length of the stock to be used. I have been reliably informed that the rig closely approximates the rig used in testing laboratories. **A word of caution — Do not use the sample tested as compression cracks, invisible to the naked eye, may have been created in the sample while undergoing stress.** However, the piece of stock from which the passed sample was removed may logically be assumed to be okay.

CONSTRUCTION TIPS

In building my test rig, I used 4 old ball bearings that I had on hand, one on each end to bear on the test piece and two to support the crankshaft, but these are unnecessary luxuries and the holesaw cutouts should work perfectly on the ends and the crankshaft should require none. Keep the spaces between the two pairs of mounting blocks and between the two moment arms at just enough more than ¾" so that the sample is free to move between them but is closely guided. On the section of a 1" dia. birch dowel used as the crankshaft and which must propel the test sample, drive in about twenty ⅝" x #19 nails and hammer over the heads and leave only enough above the dowel surface to grip the sample and pull it along. For the moment arm, hinges, I used simple steel mending plates 4" x ¾" x 1/16", obtainable at any hardware store, with one screw each into the centerline of the bearing blocks. Just below the crank and two screws each into the moment arms.

TESTING

Using the formula presented by Don Cookman but adapted for use with a sample of ¾" x ¾" X-section, a weight of 15.6 lbs. is indicated. Since water weighs 8.3 lbs./gal., I find this makes a most convenient weight and 1 gal., 3½ qts. comes out almost perfectly — however don't forget to reckon with the weight of the moment arm material.

If you wish to test a sample of Sitka Spruce of the same dimensions, a weight of 12.1 lbs. is required. I believe this is a desirable practice as I have heard of questionable stock marked as certified, also some questionable practices used in the process of certification. So why not test it yourself? The peace of mind gained is well worth the trouble.

In selecting wood from planed stock, examine the faces very carefully for evidences of hidden, or tangential, grain. The symptoms of this will be small pits on one side where the planer blades have cut against the hidden grain and gouged out small bits. On the other side, where the planer blades have cut along the grain, very fine lines at right angles to the annular rings are visible under close inspection. Wood with this hidden, or tangential, grain will have almost no strength although the apparent grain, as indicated by the annular growth rings shows beautifully straight, close and vertical and one would swear he had a very good piece of stock. I demonstrated this recently to a friend by snapping a piece in two with just the weight of the wood held in my hands at one end and had the annular rings close, vertical and straight. I would recommend fir ladder stock now as a far better source of wood as this is quite closely grade.

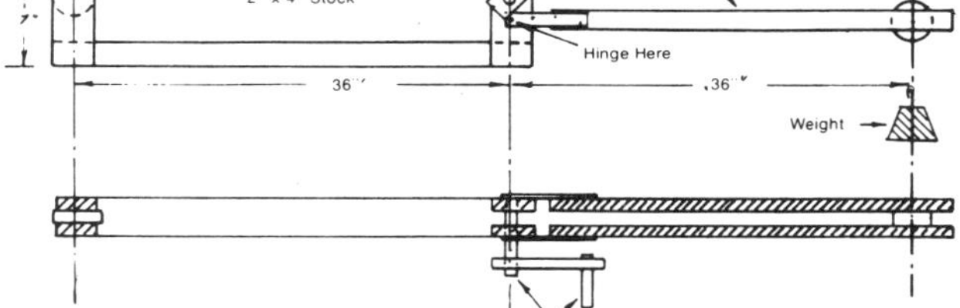

Wood and Wings

(Courtesy "FAA Aviation News")

WOOD HAS A LONG and respectable tradition in aircraft construction, going back to the gliders that preceded the earliest attempts at flight. Nor is the day of wood construction past — at least one United States' manufacturer is currently producing a line of airplanes with wooden wings. The wood spars and ribs are covered with plywood which, in turn, is sheathed with a synthetic fabric, producing a strong, lightweight wing and an extremely smooth airfoil.

Considerable numbers of wood and fabric aircraft are still being produced in Europe, notably England, France, and Germany.

There are also nearly 3000 amateur-built airplanes, most of which were made largely of wood and fabric, in active flying status in the United States. Their number grows at the rate of about 300 a year. The number of older factory-built partially wood planes still navigating the airways is not precisely known, but it is probably well up in the thousands.

The twilight of the age of wooden airplanes began at the close of World War I, when the Germans first introduced mass-produced aircraft with tubular steel fuselages. The Germans also developed two all metal airplanes in World War I. One was the Junkers J.1 which was all steel, including the skin, and was so heavy it barely flew. Only the prototype was built. More successful was the Junkers J.4, the first all metal airplane used in actual warfare. Heavily armored, it was used for ground attack. There is no record of any J.4 being shot down.

Twenty-five years later, in World War II, a reversal took place. Desperate for aluminum, the Germans developed the 522-mph jet-propelled Heinkel He-162 which had all-wood wings. Also made of wood were the wings of the rocket-powered German BP-20 which could reach 600 mph at 16,000 ft.

The Allies in World War II were not without their own wooden aircraft, probably the most famous being the plywood de Havilland "Mosquito" which was in the 400-mph class and served as a fighter-bomber, escort fighter, and unarmed photo-reconnaissance plane. The United States produced the Beechcraft AT-10 and Fairchild AT-21, both all-wood trainers. The AT-10 had wooden gas tanks, made leak proof by a plastic film lining. Vultee Aircraft produced thousands of BT-13 trainers, the aft section of the fuselage being made of wood, and the twin engine Cessna AT-8, AT-17, and UC-78 had wood wings.

LARGEST ALL-WOOD AIRCRAFT

World War II also produced the "Spruce Goose," Howard Hughes' fantastic flying boat made entirely of plywood. Originally planned as an all-metal airplane, the anticipated aluminum shortage forced designers to turn to wood. Known originally as the "Hercules" and later as the H-4, the aircraft has (it is still in existence at Long Beach, California) a wing span of 320 ft. and a hull 220 ft. long. The "Goose" weighs 200 tons plus, and is powered by eight engines. It is the biggest wood aircraft ever built. Aside from a lift-off in taxi tests, the H-4 has never flown.

On a weight-to-strength ratio, wood compares very favorably with steel, duraluminum and magnesium, the

Wood Used in Aircraft Construction

The woods listed below are those used for structural purposes. For interior trim any of the "decorative" woods, such as walnut, birds-eye maple, etc., may be used, but strength is usually of no consequence under such conditions.

ASH — Commercial White. A strong, white, flexible, coarse grained wood. Used for wing tip bow, blocking, float or hull members and any bent members.

BASSWOOD — White, soft for a "hardwood", easily worked. Used chiefly as core of plywood which has hardwood faces, sometimes for blockings, seldom as a structural member.

BIRCH — Yellow - white, hard, strong, does not split easily. Used widely for plywood and to some extent for blocking where bolts pierce a beam, etc.

CEDAR — Port Orford. White, straight grained, easily worked, very stiff and strong. Used sometimes for wing beams and compression members, for hull and float members.

DOUGLAS FIR — Yellow-white, very similar to spruce and often used in place of it, especially for ribs and small members.

MAHOGANY — Brown, coarse grained, soft for a "hardwood". Used chiefly for plywood and for inside trim.

MAPLE — Satiny yellow, very hard. Used for bearing blocks, such as is beams where bolts pierce them.

PINE—Western white or Oregon. Very similar to spruce and used for same purpose.

SPRUCE — Yellow - white, straight grained and satiny, if good quality. Used in preference to all other woods for wing beams and other structural members.

metals most commonly used in aircraft construction. Wood is less efficient by only a few percentage points. This requires a bit of explanation. By itself, a piece of wood is no match for metal, but when assembled into a structure — stringers, longerons, formers, ribs, spars — to form a unit, the difference in strength diminishes and wood is almost identical to metal in strength.

Wood is a desirable aircraft material because of the ease with which it can be worked. No expensive hydropresses, routers, riveters, drop hammers, etc., are needed. Simple carpenter hand tools, plus homemade jigs and fixtures, are all that are needed to turn out a fully certificatable aircraft. Field repairs present no serious problems.

The quality of wood used in aircraft construction is established by the FAA and other government agencies, with characteristics and specifications detailed in Advisory Circular AC 43.13-1, "Acceptable Methods, Techniques and Practices: Aircraft Inspection and Repair" (Available for $3.00 from the Superintendent of Documents, Government Printing Office, in Washington, D. C. 20402).

Wood used in aircraft structures must meet standards on moisture content, grain characteristics, type and number of knots for a given dimension, pitch pockets, mineral streaks, checks, shakes, and splints. Determining what is acceptable is a job for an expert. Aircraft quality wood is so marked.

Spruce became the wood of choice among aircraft builders because it provides the optimum strength-to-weight ratio, compared to other woods. Noble Fir, Douglas Fir, Western Hemlock, and White Cedar are stronger than spruce but heavier. Nearly matching spruce's weight/strength ratio is Northern White Pine and Yellow Poplar, both widely used in aircraft construction.

The raw lumber is used in three forms in aircraft construction — solid wood, plywood, and laminated wood. In some cases, pressure and chemical-treated wood is also used for specialized purposes.

With the exception of a relatively few items of hardware, wood airplanes are almost entirely glued or bonded together. During the assembly, thousands of small nails and brads are used, but these can be removed after the glue hardens, at no sacrifice of strength.

Aircraft glue is not to be confused with common fish and hide glues which are totally unsuitable for aircraft fabrication. These deteriorate rapidly, have low strength, and are difficult to apply.

Aircraft glue development has come a long way from the casein adhesives used in pre-World War I and subsequent aircraft up until the early 1940's. While casin is still used, it is gradually being edged out by resin glues which are not only stronger, but are absolutely waterproof and impervious to attacks by mold and fungus.

Among the newer glues are resorcinal-formaldehyde, urea-formaldehyde, and "fortified" urea-resin. They are not limited to aircraft manufacture but they have a wide application in the wood fabrication industry, and are readily available.

GOOD DESIGN VITAL

Proper design, allowing for free ventilation, drainage, and access to points where moisture is likely to accumulate, is as important as the craftsmanship and materials going into the aircraft. Inspection plates and "peep holes" should be provided so that interior examination of spars, bulkheads and compartments can easily be made.

Wood is vulnerable to concealed damage by wood-eating insects, moisture, and microscopic fungi spores. Insects reveal their presence by tiny holes and minute deposits of chewed wood — sawdust. Some insects can actually be **heard** as they go about their destructive work. Gentle probing with a dull knife point around suspected areas will reveal the extent of infestation and consequently the amount of destruction. Wood properly treated with a sealer such as varnish or a resin mixture during the aircraft assembly process, or following repairs, is protection against insect damage.

Wood that appears to be perfectly dry sometimes crumbles to the touch or light pressure. Since there is no evidence of moisture, the assumption is natural that some mysterious process, not related to moisture, has destroyed the wood. In point of fact, rot can set in even in the absence of moisture. Rot is caused by fungi.

Fungi spores are wafted through the air in a continuous, invisible stream, everywhere. They settle on everything, including wood. The spores can remain dormant for extended periods of time — one, five, or ten years — and come to life when conditions are right.

The most suitable environment for the spores to germinate is about 70 degrees F, when the wood is damp. For growth the fungus needs a mild temperature. The resurgent spores penetrate the wood with vein-like roots which feed off the wood's cellulose, leading to its destruction.

FUNGUS IS DESTRUCTIVE

The deterioration is subtle and can remain hidden until disaster overtakes the plane unless regular, careful inspections are made. As the fungus is developing, the infested wood darkens and there is evidence of warping, shrinkage, and cracking taking place.

In the final stages, the fungus "fruits" as a mushroom, puffball, or the leathery growths seen on dead trees.

It has been estimated that a mushroom three inches in diameter may give off one to two **billion** spores, each capable of starting the destructive cycle all over again.

Paint and varnish will not prevent rot, nor will it "cure" it, since paint and varnish deteriorate with age. Indeed, the vegetable oils in some paints provide nourishment for fungus growth.

The most effective way to eliminate rot is to keep the wood dry — the moisture content should not exceed 20 percent. Moisture content for aircraft-grade wood is 12 to 15 percent. This degree of dryness is assured by the quality controls placed on aircraft-grade woods, but the day-to-day moisture content found in the wood members of an aircraft changes with the weather.

Moisture also penetrates the aircraft in the form of rain or snow entering through poorly fitted doors, stabilizer slots in the fuselage, and through wing strut and aileron holes. It is spun into wheel wells when operating from wet or snow-covered runways or muddy fields. Condensation forms in poorly ventilated places, creating an alternately wet and dry atmosphere ideal for rot to develop. Plugged or non-existent drain holes provide standing reservoirs of water.

Unless a pilot is a certificated airframe mechanic he is prohibited from doing any repair work, however minor, on wood structures other than amateur-built aircraft. His best defense against rot and insect damage is vigilance — and taking care of his airplane.

A wood airplane should be hangared, more so than a fabric-covered metal-framed aircraft. Even though the wood is treated with several coats of varnish, time and the action of airborne chemicals cause the varnish to deteriorate. Hair-line cracks admit moisture, opening the door to rot. Metal fittings in contact with wood invite rot, and these should be carefully inspected for rust stains.

Whenever rotting is suspected, or when there is a suspicious bulge or evidence of internal damage due to impact, a qualified mechanic should be consulted. If there is damage, it won't go away by itself, and waiting for the next inspection might be too late.

Wood remains an excellent aircraft material. How well it performs depends on sound design and conscientious maintenance. The latter begins with the pilot or owner.

Aircraft Construction Materials

It has been found by laboratory tests that the strength of wood along the grains averages about 17 times as great as the strength across the grains. This fact makes it highly impractical to use veneer (a thin single sheet of wood) in aircraft construction, as either an excess of weight or a lack of strength results. To overcome the disadvantage of this variation of strength, plywood has been developed. This was done by gluing two or more sheets of veneer together with the grains running in opposite directions. The more plies, or sheets of veneer, the more constant is the strength ratio of the finished product. As can be readily seen, the strength of plywood depends upon the selection of the material for the plies and the characteristics of the glue.

After much experimentation with different materials, birch, mahogany and spruce were found to be the best for the various purposes. The glue presented a larger problem than the selection of material. At present three types of glue seem to answer the need for a strong, water-resistant glue; the blood albumin, casein and phenolic glues. With these glues it is possible to develop a glue bond capable of withstanding a shear stress of 350 pounds per square inch and over when it is dry and as high as 250 pounds when wet. By using carefully selected veneer and the proper glue a product has been made that possesses properties of weight, hardness, strength and elasticity in practically constant ratios. This makes aircraft plywood truly an engineered material.

Aircraft plywood is furnished in many thicknesses, with a varying number of plies, the most popular size range being from 3/64" to ¼" and from 3 to 5 plies. However, for special purposes it is furnished in standard sizes down to 1/40" in thickness, such as wood used in gliders, and can be had in standard sizes up to 1" for structural parts. By special order plywood can be purchased in any size from .030" - 3-ply, to 2" - 20-ply or over.

In the manufacture of plywood it has been found advantageous to make the ply that is in between two hardwood plies of a soft wood, such as poplar or basswood. While this inside, or core ply, does not add much to the strength of the plywood, it does provide a stronger glue bond than if all hardwood were used. For this reason plywood is usually furnished in 3, 5, 7, 9, etc. plies, which allows a hardwood ply on each face with a soft wood filler or core ply, between each of the hardwood sheets.

Balsa wood is also used as a core ply when soundproofing plywood is made. This is used extensively in cabin linings as it serves both for insulating and soundproofing the cabin. This plywood is supplied in a variety of hardwood faces suitable for trim.

Plymetal is made by cementing an aluminum or dural face to a stiff plywood panel. This makes an exceedingly strong product which is approximately 50 times as stiff and has 6 to 8 times the impact resistance of sheet metal of the same weight. Plymetal is used for floorboards and other such places as demand a material which will withstand rough usage.

Of the types of glue used as a bonding agent in aircraft construction, casein glue is the most common. This glue is made from dehydrated milk curds which have been ground to a white or cream colored powder. It is used almost exclusively in aircraft construction due to its durability, convenience and economy. It was first introduced in the early years of World War I and proved its worth in the manufacture of water-resistant plywood for the airplanes of that period.

While the strength of a glue bond of any kind of glue depends upon the material and the fit of the surfaces, the pressure applied, atmospheric conditions at the time of drying, etc., under ideal conditions casein glue develops a bond capable of withstanding a shear stress well over 250 pounds per square inch. Although this glue is often called waterproof, this is not strictly correct. It is, rather, water resistant and the strength of this glue in a wet shear test is still over 125 pounds per square inch. The wet shear test was made after specimens had been soaked in cold water for 48 hours.

Casein is a cold water glue, meaning that it is mixed for use with cold water (70 degrees Fahrenheit) and can be applied to cold lumber with equally good effects. These two items add greatly to the convenience of this glue. The fact that a very small quantity of glue can be mixed at one time, and that it does not have to be used immediately, and that the unmixed powder can be kept over a long period of time, if kept away from moisture, makes casein glue a very economical glue for the mechanic or the factory. The time required for drying is somewhat greater than that of most other glues, taking approximately 24 to 48 hours to dry completely. However in most cases it is safe to work glued surfaces and joints after the material has been in the clamps for 12 hours.

Here are some helpful hints in using casein glue:

1. Never use casein glue after it has been mixed for over four hours.

2. For gluing cracks in wood, use a thinner mixture which will run in and fill up the cracks.

3. Prepare all necessary clamps before applying the glue.

4. Where possible, put paper between the material and the clamping block to prevent the glue sticking the work to the blocks.

5. Make sure the surface to be glued is clean, free from all grease paint, varnish, dirt, etc.

6. Glued joints must fit perfectly.

7. Wipe off all excess glue after clamping. When the glue dries, it is extremely hard on cutting tools and it also has the tendency to attract moisture even through varnish, which will cause the material to rot under the exposed glue.

8. A pressure of 175 pounds per square inch is needed when glu-

ing spruce to get the best results. Slightly higher pressure is needed for the hard woods.

9. Glued material should never be removed from the clamps under 12 hours.

10. A small, soft brush is helpful in applying glue.

11. Never sandpaper surfaces to be glued, as the sandpaper dust clogs the pores of the wood.

12. Casein glue can be used to cement fabric to wood if the sizing is first removed from the fabric by rinsing in cold water.

Nail Positioning

by Joe Stephens

Recently while spreading adhesive for covering a flooring area with asphalt tile, I noticed that the serrations on the trowel were cut in increments of 3/16 and 1/8 inch pitches. Wanting my F-51 to show good workmanship, it occurred to me that the trowel would make an excellent guage for laying out a uniform nail pattern for wood and a satisfactory guage for many rivet patterns.

I found that the best procedure was to place a piece of masking tape alongside and parallel to the upper serrated side of the trowel and mark the desired increments of spacing. Draw a line on the workpiece where the nails

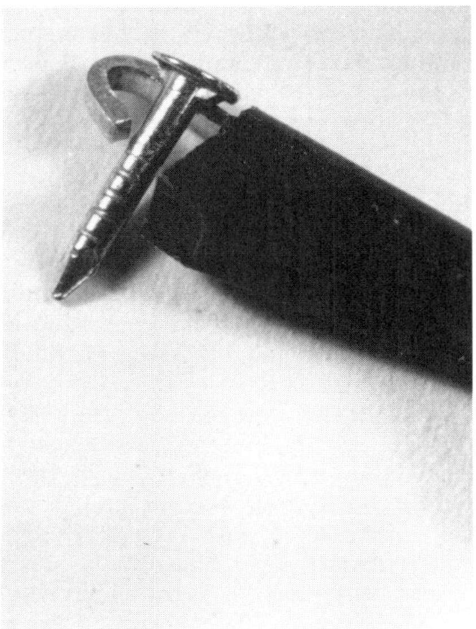

are to be driven and drive the first nail approximately halfway into the workpiece. Place the trowel with the first index serration around the nail and then drive a nail halfway into the workpiece at the last index mark on the trowel. Using the first and last nail positions for positioning the trowel, clamp it in place and after piercing the remaining index holes with a sharp pointed instrument, drive the nails into the holes about halfway. Remove the trowel and finish driving the nails, except the final nail which becomes an index point for the next course of nails. The degree of uniformity thus obtained should be satisfactory for laying out a suitable pattern for construction work where nails are to be driven permanently into plywood webbing or sheathing or to lay out an accurate pattern for nailing strips.

I also found a spring loaded test lead hook type probe to be great for picking up small nails. The head of the probe is depressed to expose the hook which is used to pick up the nail and place it in position for driving in place. After setting the nail about halfway into the workpiece, release the nail by depressing the head of the probe and finish driving the nail. The hook probe can be purchased at most electronic supply houses or from Radio Shack stores. These hook-type probes are very popular among TV and computer technicians and are relatively inexpensive.

Cutting Patterns / *by Randy W. Carrels*

If you have gone through the process of cutting a pattern for wing ribs or other multiple parts, cutting a master can take up some precious time.

I simply cut out the pattern from my plans sheet, then apply rubber cement and put it on a blank (plywood larger than pattern size). Clamping multiple sheets of plywood together for ribs (in my case the KR-2), I rough cut the plywood to size, then finished the parts on a belt sander using the paper pattern as a guide. The best part is peeling the pattern off and lightly rubbing any residue away, leaving perfectly matched parts.

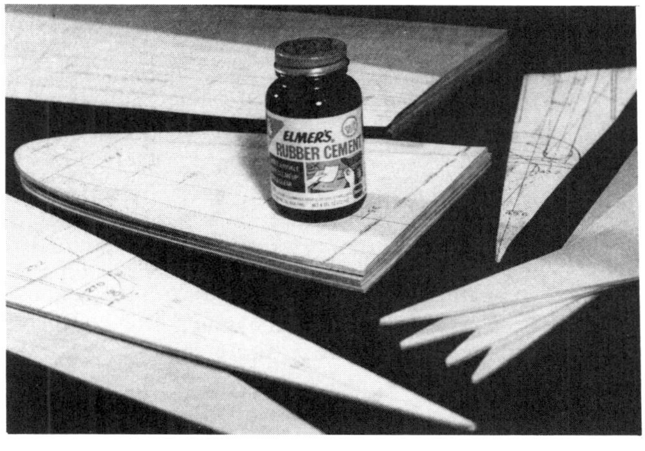

For the Wood Butcher

By Ray Borst, EAA 1526

Additional information on sanding for glue joints, discussed by Bob Whittier in his article "More on Gluing" which appeared in the April issue of SPORT AVIATION, can be found in Report No. 1351, "Improving the Gluing Characteristics of Plywood Surfaces by Sanding", put out by Forest Products Laboratory, USDA Forest Service. I have prepared à summary of this information extracted from the report which gives the gist of the laboratory findings as follows:

"The condition of wood surfaces at the time of gluing has an important influence on the strength of glue joints . . . Fresh wood surfaces that are smooth, uniform, and free from crushed fibers are generally most satisfactory for gluing purposes . . . Sanding, tooth planing, or other methods of roughening well-planed wood surfaces are unnecessary and inadvisable for fresh surfaces.

"In contrast to well-planed normal wood surfaces, plywood sometimes presents more difficult gluing problems . . . Surface conditions that may develop during veneer and plywood manufacture are believed responsible for these occasional difficulties . . . Light sanding of the plywood surfaces to be glued is the most effective and practical means thus far found for correcting such unfavorable surface conditions . . . In investigating the efficacy of light sanding for improving the gluing characteristics of plywood, a number of studies were made . . . One-half of the plywood strips were lightly sanded with 3-0 garnet paper before gluing and the rest were left in their original condition . . .. The results of these tests are summarized in Table 1. They show that, in every instance, light sanding effected a material improvement in the quality of the glue joints.

Table 1.—Results of shear tests on sanded and unsanded plywood nail-glued to Sitka spruce cap strips with a cold-setting, urea resin glue. [1]

Plywood Species	Appearance of surfaces	Surface Treatment	Breaking[2] Strength	Wood Failure	Percent Increase of Strength
			Pounds per square inch	Percent	
Yellow birch	Normal	None	364	28	
Do	do	Light sanding	520	99	43%
Sweetgum (heartwood)	do	None	340	45	
Do	do	Light sanding	628	98	85%
Sweetgum (heartwood)	do	None	420	63	
Do	do	Light sanding	500	100	19%
Mahogany	Heavy bleed-through	None	389	30	
Do	do	Light sanding	519	99	34%
Mahogany	Wax on the faces	None	186	0	
Do	do	Light sanding	445	53	140%

[1] Comparisons are between sanded and unsanded samples of the same lot and not between species or lots because the plywood panels were of different thicknesses and constructions.

[2] Each average is for 9 or more shear specimens cut from 3 or more nail-glued assemblies of the type shown in Figure 1.

"Light sanding is obviously hard to identify. The danger of over-sanding and consequent reduction in strength has been the principal objection to sanding aircraft plywood . . . Three types of sanding, designated as light, medium, and heavy, were used. Light sanding consisted of 10 single strokes across the face ply, medium sanding of 20 strokes, and heavy sanding 40 strokes. The reductions in face-ply thickness are shown in Table 2.

"Light sanding did not reduce face-ply thickness by more than .002 inch, or 10 per cent . . . Medium and heavy sanding, on the other hand, produced greater reductions in thickness, which were possibly sufficient to cause significant strength decreases.

Table 2.—Reduction in face-ply thickness produced by sanding 0.020-inch faces of plywood with 3-0 garnet paper [1]

Species	Reduction in Face-Ply Thickness					
	Light sanding (10 strokes)		Medium sanding (20 strokes)		Heavy sanding (40 strokes)	
	Inch	Percent	Inch	Percent	Inch	Percent
Yellow birch	0.0005	4	0.002	10	0.004	20
Sweetgum	.001	5	.002	10	.006	30
Yellow-poplar	.002	10	.004	20	.007	35
Douglas-fir	.002	10	.004	20	.005	25

[1] Each value is an average of thickness measurements on 5 plywood strips of the type used in nail-gluing tests (Fig. 1). Sanding was done by hand, across the grain.

"Another test was made with different grades of garnet paper, and the reductions in thickness caused by each were measured. The face plies were lightly sanded across the grain by 10 strokes. The reductions in thickness of face plies produced by three types of garnet paper are shown in Table 3. It is evident from these results that the danger of over-sanding and seriously reducing face-ply thickness decreases with the finer grades of paper. The thickness reductions with 4-0 garnet paper were hard to detect for birch and averaged about 5 percent for the softer species . . . In all the sanding tests reported here, the sander dust was removed with a clean cloth or brush.

Light hand-sanding of plywood surfaces has been found very beneficial in correcting surface conditions that interfere with the adhesion of cold-setting glues in aircraft-assembly operations. Light sanding should not be considered a cure-all but a useful aid in correcting the gluing characteristics of plywood otherwise difficult to glue. The danger of over-sanding can be reduced by specifying the use of sandpaper no coarser than 3-0 garnet on species of high density and no coarser than 4-0 garnet on species of low density. Reductions in face-ply thickness should never exceed 10 percent and whenever possible should be limited to .001 inch. Frequent thickness measurements are advisable when hand sanding is practiced. In order to be of the maximum value, light sanding should follow masking and sealing operations."

Forest Products Laboratory has many other valuable reports and other publications on wood available. Ask for List RPI-4, Publications on Structural Sandwich, Plastic Laminates, and Wood-Base Aircraft Components. They also have a supply of ANC-18, Design of Wood Aircraft Structures. Write to the Director, Forest Products Laboratory, Madison 5, Wisconsin.

Table 3.—Reduction in face-ply thickness produced by lightly sanding 0.020-inch faces of plywood with different grades of garnet paper[1]

Species	Grade of Garnet Paper					
	4-0		3-0		2-0	
	Inch	Percent	Inch	Percent	Inch	Percent
Yellow birch	0.00	0	0.001	5	0.002	10
Sweetgum	.001	5	.002	10	.003	15
Yellow-poplar	.001	5	.002	10	.003	15
Douglas-fir	.001	5	.001	5	.004	20

[1] Each value is an average of thickness measurements on 5 plywood strips of the type used in nail-gluing tests (Fig. 1). Sanding was done by hand, across the grain.

Glues and Gluing

By Jack Millington, EAA 4101

The use of glue to produce strong, reliable joints between wooden members is a well-established practice. Unfortunately, the popular concept of glue is based on experience with cheap furniture and makeshift repair work and this has led to a lack of appreciation of the results which can be obtained through the use of good glue and careful workmanship.

A good glue joint in wood is akin to a good welded joint in steel. Both are capable of developing the full strength of the material of which the structure is made . . . but to do this the joints have to be properly designed. In the case of glue this means having a working knowledge of the properties of glue, how it "sticks", how it behaves under tension loads and shearing loads, the effect of wood grain on adhesion, the effect of structural flexing on joints, and so on.

The failure of properly made glue joints invariably occurs in the wood, not in the glue line. Inspection of a failed glue joint will show how the wood fibers have been torn. The strength of a glued joint is not a result of the "stickiness" of the glue, or of the glue acting as a connecting link between the pieces of wood. Instead, in a good joint the wood faces are in actual and uniform contact and the glue bonds the pieces together chemically through the penetration of the glue **into** the surfaces of the mating parts.

Nails, screws and staples are frequently used in glue work and it is important to understand the part they play. They have nothing to do with the strength of the final joint but are used only to apply pressure while the glue is setting. After the glue has set it is quite possible to remove them without materially reducing the strength of the joint. In fact, when plywood skins are used over wooden wing and fuselage frames, the nails are often driven through "nailing strips", or long thin battens of wood. The nailing strip helps distribute pressure, and when the time comes it helps pull out the many small nails. When feasible, nails may be removed. However, if screws are used their removal leaves holes of appreciable size which can detract from the wood's strength. Thus it is usual to leave screws in place, as they serve to fill the holes they have made. Incidentally, the FAA does not approve the used of wood screws in the primary structure of airplanes.

Modern waterproof glues are not subject to deterioration from moisture, oil, fungus attack, etc. It is important to remember that there is a difference between a glue that is described as water-**resistant** and water**proof.** Some early glues based on animal products were so palatable to fungus that cases are on record where plywood came apart by reason of the fungus working its way through the glue lines! Modern glues are capable of resisting forces of up to 1½ tons shear stress per square inch.

Each form of material used in construction work demands its own special design techniques. A bridge over a river might be built of steel, masonry, wood or concrete. Four bridges, one made of each of these materials, would all serve the same purpose, but each would reveal a distinct design technique which would be based on the material used. Similarly, glued spruce and plywood aircraft structures follow design practices which differ from those used for metal construction. There are two all-important principles for wood:

1. Glue joints must be loaded in shear only.
2. The grain of the wood must lie in a suitable plane.

To be glued successfully, wood must have the proper moisture content; about 12 percent is ideal for aircraft work. It is wise to glue members together on the same day they were faced up, to avoid even the slightest warpage of mating surfaces brought about by changes in moisture content.

Joint faces should be smooth, free of planer marks, and contain no splinters or loose material of any kind. While good surfaces may be obtained by careful use of a hollow-ground novelty circular saw, the best faces for small areas are produced by careful hand planing. Serrations on the surface caused by nicks or teeth on tools prevent intimate contact of mating faces and are to be avoided. Sandpapering should also be avoided as it is apt to destroy the true flatness of the work and will clog the wood pores with wood dust and abrasive grit, which are unfavorable to good bonding. Some of the skilled old-time airplane mechanics used scraps of window glass for the final scraping job. They would use a straightedge and glass cutter to make a perfectly straight cut on some glass, crack the glass on the score line, and use the pieces as scrapers, holding them so that one edge of the glass would be drawn over the wood at such an angle as to cut a very thin layer off the surface. They also used blue carpenter's chalk to check on fit. Chalk would be rubbed on one part of a splice and the two pieces pressed carefully together. Any high spots would

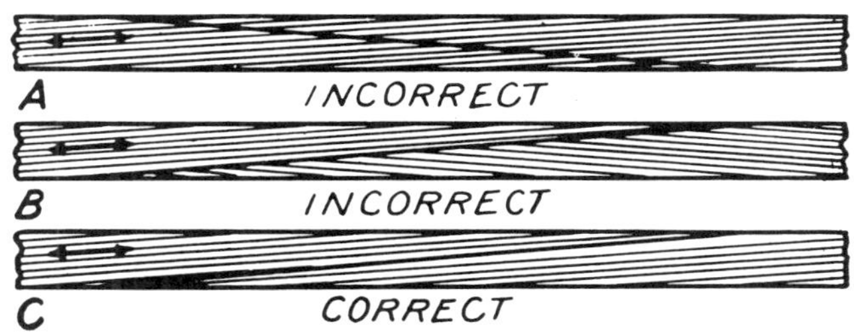

These examples show how direction of grain in glued members has a great effect on strength of the final joint. In A and B there is more or less an edge-grain contact which is weak.

of course show up where chalk had rubbed off. A final light scraping with glass would remove all traces of chalk dust.

Cleanliness is of paramount importance. It is impossible for glue to bond to surfaces contaminated with oil or wax. Always use a lead pencil for marking, instead of a wax crayon. In repair work all varnish, paint and old glue must be removed by scraping and planing down to bare wood. Remember, some paint removing fluids contain wax.

Fitting difficulties during gluing-up operations can be avoided by assembling pieces without glue beforehand and inspecting for close fit all around. When several members are being joined at one time, such as the capstrips and verticals of a truss rib, it is essential that the edges of the different parts be level with one another to present a flush surface to the gusset. Similarly, in building a box spar the flanges and spacer blocks should be made first, allowed a day or two to stabilize, and the edges of mating pieces dressed flush with each other just before applying the plywood webs. When ready to glue, have all needed items at hand to make assembly time as short as possible.

It should be emphasized that the common fish glues and hide glues are quite unsuitable for aircraft work. They are difficult to apply properly, have low strength, and deteriorate rapidly. At the same time, modern production-type resin adhesives requiring high curing temperatures cannot conveniently be used by the amateur. A suitable glue must be easy to use, of very high strength, highly resistant to deterioration, and readily available.

Casein glue has been used in aircraft work for over 40 years and urea resin glue for about 15 years. The more recently developed resorcinol resin type is so superior to any others, however, that it is now the only kind to be recommended. Casein glue is comparatively cheap, but as a typical amateur airplane does not use a great amount of glue this is of small significance. It can be used at any temperature above freezing but resorcinol glues will not set at temperatures below 70 degrees F. This is normally no disadvantage. Resorcinol glue is less sensitive to variation in moisture content of the wood and does not require such a high clamping pressure as urea resin glue. Because it is not prepared by mixing with water it does not change the moisture content of the wood to any appreciable extent . . . an effect which can be significant with casein or urea glues. Resorcinol glue is absolutely waterproof and impervious to mould or fungus, while the others, especially casein glue made without a preservative, are only water-resistant and deteriorate under conditions favorable to fungus growth. Joints made with resorcinol glue maintain their original strength after prolonged exposure to severe variations in temperature and humidity, where similar conditions would make casein or urea glued joints fail completely.

A resorcinol glue readily available to the amateur and eminently suitable for aircraft work is "Elmer's Waterproof Glue", sold at hardware stores. This was formerly called "Cascophen". It is packaged in convenient-size containers, meets government specifications for this type of product, and is the same material as supplied to aircraft factories. It may be used at any temperature above 70 degrees F. The following recommendations are applicable to this type.

The glue is a thermo-setting resin which hardens through a chemical reaction rather than by "drying out". It consists of two ingredients, a reddish, transparent liquid resin and a finely-powdered catalyst or hardener. These are mixed just before use and when they set cannot be dissolved, or be softened by heat.

Containers for mixing the glue should not be of copper or brass. An old glass food jar makes an ideal container for small lots. The quantities of glue ingredients should preferably be weighed, but may be measured by volume. Use one part by weight of powder to four parts by weight of liquid, or three measures of powder to four measures of liquid. Too much powder will make the glue too thick to use, and too little will result in a weak joint. Add the powder to the liquid gradually, stirring constantly to break up lumps and work out air bubbles. Stir the mixture, do not beat it. Never use a high speed mixer. As soon as blending is complete the glue is ready for use.

At a room temperature of about 70 deg. F. the glue will remain usable for about four hours. At 95 deg., the pot life will be shortened to about an hour. On a hot day the pot life may be extended by placing the glue container into a bath of cool water . . . about 70 deg. is suitable. This will help hold down the glue's temperature. Mix only as much glue as is needed for the job at hand. Discard the glue when it begins to thicken.

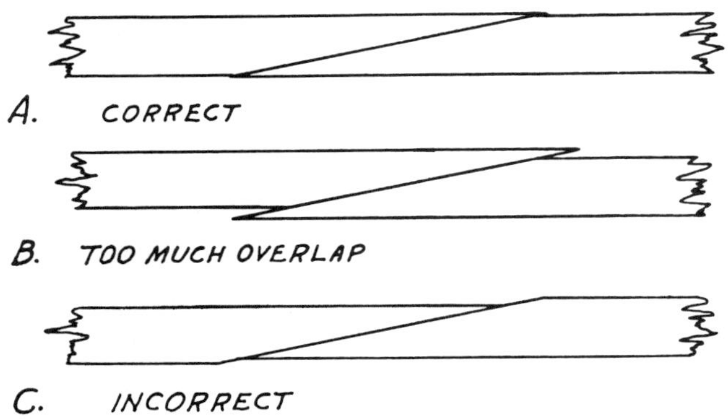

Care is needed in fitting joints for aircraft work. These three examples show how easy it is for an ostensibly simple, foolproof joint to go together wrong. In B and C, a plywood gusset glued to outside of these splices would have poor contact with the wood and the resulting pockets of solid glue would be much weaker than if a good wood-to-wood bond had been obtained such as in A. Often, aircraft nails are used to prevent slipping or shifting of parts as pressure is applied.

Glue may be applied with a wooden spreader stick for small parts or with a brush for larger surfaces. One of the "acid swabs" sold in hardware stores, a little stiff-bristled brush with metal handle, does a nice job on moderate-sized work and costs little. Under ideal conditions only one surface need be covered with glue, but more often the glue is applied to both mating faces. This is particularly important in scarf joints. One coat may be forced into the pores of the wood with a putty knife and be followed by a brushed coat on each piece. It is important to spread glue uniformly.

As little time as possible should pass between spreading the glue, assembling the parts, and applying pressure. Open assembly time (before parts are put together) is most critical and must be kept short . . . about 15 min. is the allowable time. Maximum closed time (between assembling parts and applying pressure) should not exceed 45 min. These times cannot be added together; if 5 min. are used for open assembly it is one-third of the allowable 15 min. and you then have two-thirds of the closed time or 30 min. in which to apply pressure. Usually we never approach these limits but they should be kept in mind in case things should go wrong and there is some question of how long it will take to set them right before proceeding.

Thin plywood clamped by small nails is more sensitive to time of assembly than are heavier sections which can be firmly clamped. Note that the times given above apply to room temperatures near 70 deg. At about 95 deg. only one-third of these times is allowable due to the accelerating effect temperature has on the chemical setting action.

If glue is evenly applied in sufficient quantity and assembly is achieved in reasonable time with uniform pressure, a uniform bead of glue will be forced out around the joint edges. This is an indication that a good joint has been achieved and should be expected on all critical parts. Excess glue should be wiped off before it hardens.

Jobs should be left under pressure for the following times:

Room Temp.	Hours
70 deg. F.	10
80	6
90	3½

A rough but useful method of maintaining a check on joint quality is to glue two small strips of wood together each day that work is done on the airplane. Use a lap joint, and leave them under pressure overnight. Shearing the pieces apart in a vise should reveal fractures in the wood fibers rather than in the glue line.

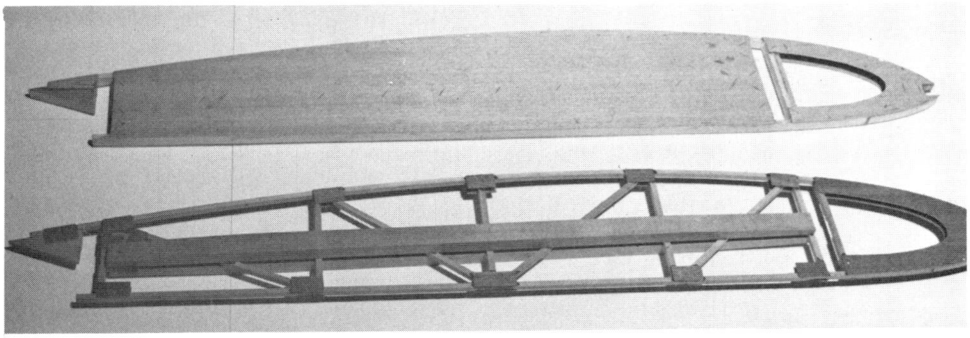

Compression ribs are usually built with plywood sheet covering both sides of the rib as shown at the top in the photo left. Another popular method is to glue reinforcing strips horizontally on both sides of a built-up rib, as shown below.

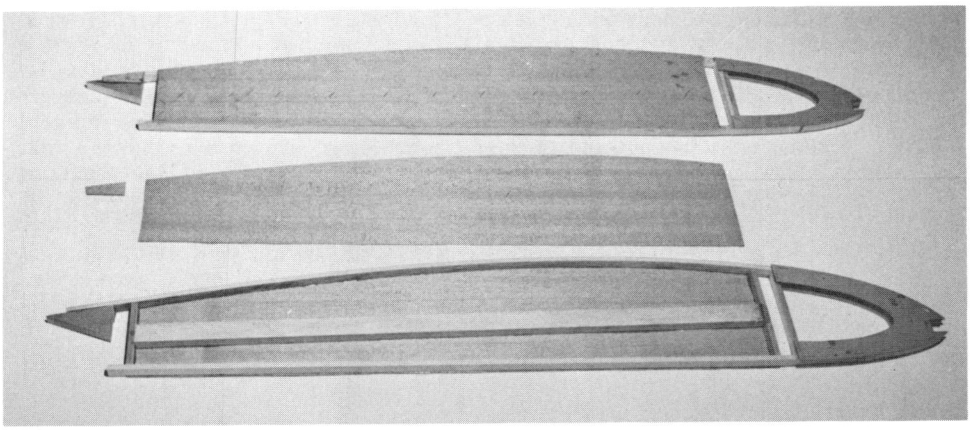

Steps in construction of the plywood sheet type compression rib also includes a single horizontal reinforcing stip sandwiched between the plywood skins. Builders use small aircraft type nails to hold parts in position while glue dries, or some builders staple parts together. When staples are used they should be removed after glue sets. These ribs will be positioned forward of the aileron.

Ribs in place and ready for gluing to spars. Gussets will be added fore and aft of the spars to hold the ribs. Drag wires have yet to be installed. When completed, wing panels are extremely strong, yet very light in weight. Wing panel shown here is from the Pober Pixie. Airfoil is the popular Clark Y — an ideal airfoil for low speed aircraft.

Making Cantilever Wing Spars

by Terence W. Algeo

DURING THE summer of 1967, two Piel "Emeraude" wing spars were built by Vern Glass and myself, using a jig and procedure which I shall describe here for the benefit of others who might wonder how to go about this part of the task of building a wooden cantilever wing.

The basic procedure was first to make a good, accurate workbench, then attach to its working surface suitable locating blocks for some curved stringers against which the spar material could be clamped and glued to accurate shape. The spar was made of upper and lower "booms" or internal beams, separated by vertical spacer blocks and the whole boxed in with plywood faces.

The bench top was made of ¾-in. commercial grade plywood, ripped into 2 x 8-ft. strips. Legs made of 2 x 2-in. material were set at 4-ft. intervals. One-by-three-inch material was used for crosspieces, diagonals and stringers to tie things together rigidly and support the plywood against flexing. Enough weights were added under the bench to make a rigid and steady work support of it. The general setup is shown in a drawing accompanying this piece.

Use was made of a length of plastic tube filled with colored water to level this bench longitudinally, while a carpenter's level served the same purpose laterally. We used shims under the legs to level up the table, but if we were doing it again we'd consider it worthwhile to make and use adjusting brackets and bolts as shown in the workbench sketch, because of the greater ease of obtaining and holding the kind of fine adjustment wanted.

We used a tight wire as a straightedge from which to lay down an accurate reference line on the bench top. We found that a fine-pointed felt marking pen was excellent for making legible and durable lines on the plywood. Working from this reference line, we plotted the shape of the spar on the surface of the bench. The accuracy of the bench's surface took care of trueness of the spar in regard to warpage, and care in plotting took care of accurate wing dihedral. You can't end up accurate if you don't start out accurate!

Taking care to make sure that the table saw was set accurately for perfect 90-degree cutoffs, we made a supply of jig blocks which are numbered "1" in the sketch. These we attached to the bench surface with wood screws and just enough glue to add rigidity, for later it would be desirable to be able to knock them off without doing much harm to the workbench's surface. These blocks were made of dry, clear wood as was the clamp stringer, numbered "2" in the sketch. The jig blocks were located as close as eight inches apart at points of maximum bend in the booms, and farther apart elsewhere as common sense dictated. We took much care to make sure the working faces of the clamp stringers were accurately aligned over the spar drawing lines on the surface of the bench.

The bench ready, we began actual work on the booms by machine-dressing to required dimensions enough aircraft-grade Sitka spruce for the several laminations. We left between ⅛ and ¼-in. surplus material along the edges of the strips where the glue would squeeze out, to allow for cleaning up the surfaces with a plane after the glue had set. We wanted fresh, clean wood-to-wood contact at every glued joint, hence our care to clean oozed glue off completely not merely by wiping but by planing off the glue-impregnated surfaces.

To get the required lengths for the booms in the "Emeraude" wing, we spliced pieces of spruce together using long-slope scarf joints. The splicing was done by laying the wood on the aforementioned reference line to get the two pieces straight before gluing. Several small nails were used in each scarf joint to prevent shifting of the wood as pressure was applied. We used clamping boards, similar to those visible in Fig. 8. Wax paper was used under the work to keep oozed glue from sticking it to the bench. Every time we used wax paper we were very careful to remove all traces of wax, lest it weaken later glue joints or interfere with adhesion of varnish. Fig. 1

Fig. 1

shows two splices before being clamped, and Fig. 2 gives an overall view of the workbench and splicing operation.

Fig. 2

When all needed laminate strips were ready, assembly of the booms began. Fig. 3 shows the lower boom

Fig. 3

clamped to the clamping stringer on the workbench. We found it necessary to use several pieces of wood on top of the work to hold the laminations down against tendency to slide and lift as the pressure was applied and glue began to flow.

We learned that the critical factor in making laminated parts of this size is glue-setting time. It is necessary to study up on the pot-life characteristics of the glue being used. On a big job there could be so many laminations and glue lines that a quick-setting glue would leave insufficient time to get everything put together and clamped before the glue began to set.

The glue we used was an epoxy type — H. B. Fuller's *Resiweld No. 7004* — mixed to the "rigid ratio" of two measures of part A to one measure of part B. This particular glue sets slowly, affords good wood penetration and, due to the slow setting, parts could be shuffled as necessary when fitting laminations together before tightening the clamps. Little pressure is required with this glue but, close-fitting joints are desirable. We clamped every four inches on the average except for some places where backing boards were used to distribute the pressure.

We left our booms on the jig for several days to allow the glue to cure thoroughly. Upon removal we dressed off the surplus wood from their sides with a plane. A power plane is shown in Fig. 4, but we

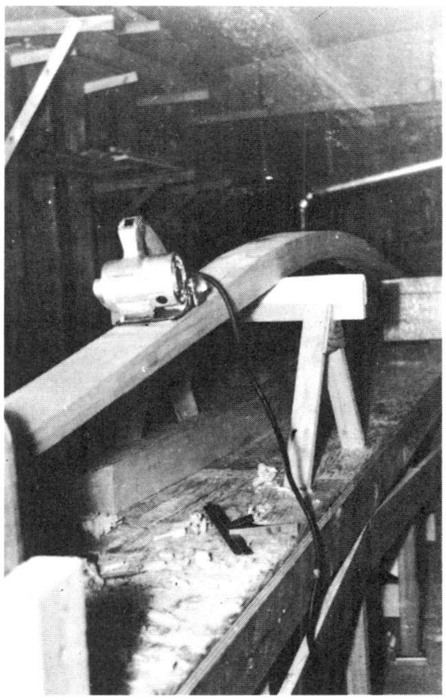

Fig. 4

found a hand plane was adequate for most of the work. We laid the upper and lower booms on the workbench close together and laid a straight-edge across them at many locations to check parallelism of their side faces. This was important as plywood webs would later fit against the faces. Some of the laminations have ends which taper off into feather edges; on this particular design the tapering is done with a plane after the booms have been glued, and that is what is going on in Fig. 4 — one of the booms has been set up on temporary horses to hold it in position for this operation.

To assemble the spars we found it expedient to remove the clamping stringer and jig blocks used to make the lower booms, leaving only the upper stringer on the bench. This made it possible to clamp the upper boom to the jig to maintain alignment, but to move the lower boom back and forth as needed to facilitate also the fitting of the numerous separator blocks between the two booms.

During this assembly work we worked from the center of the spars out to the tips. Long pipe clamps were used to pull the lower boom into correct position as each separator block was fitted. One such joint is shown in Fig. 5. Reliable gluing

Fig. 5

depends on maintaining whatever minimum air temperature may be specified on the container. In our case, we maintained the shop temperature at 70-degrees F. while the booms were being glued. When working on smaller joints during spar assembly, we allowed overall shop temperature to vary between 50 and 60 degrees F. depending on the weather, and maintained local temperature in the vicinity of joints at between 85 and 105 degrees F. by erecting cardboard hoods as in Fig. 6 and directing into them the warm air from an electric heater.

The surface of the workbench served to line up the booms and separator blocks on their lower surfaces. When all assembly work had been finished we found slight irregularities in thickness on the top surface. We used fairly coarse wood rasps to take down the higher points

Fig. 6

Fig. 7

Fig. 8

after which the whole surface was cleaned and trued up using a very sharp hand plane set for a fine cut.

The plywood web pieces were applied in several sections as determined by raw material length. Backing strips were used where plywood joints did not come over separator blocks. We used enough nails to hold the plywood sections in place against slipping when being glued and clamped. A web piece ready to go on is shown in Fig. 7, with nails started and end-beveled to fit accurately against the plywood already in place. Each piece of web plywood was cut a little wider than the spar, the surplus being trimmed off after gluing.

Pressure was applied by lengths of ¾-inch commercial plywood cut a few inches wider than the spar. Matching holes were drilled through these pieces and into the bench top. Long bolts with plywood washers and wing nuts were used to pull things together. Fig. 8 shows one of these setups; do not be confused by the spar to the upper right of it which happened to be there when the picture was taken — the plywood pressure plate has a level resting on it.

When all web sections had been glued to one side of the spar, the spar was removed and surplus plywood dressed from the edges. The spar was then reversed end for end and positioned against the gluing stringer with open side up. We drilled ventilation holes as called for in the plans and applied finish to the inside surfaces of the spar taking care not to get any of it on the wood faces still to be glued. On the bottom sides of each piece of web plywood we drew in outlines of the booms, spacer blocks, etc., and carefully applied finish to the non-glued areas only. At this point we had the spars inspected by a Department of Transport aviation agent and approved for closing up. Fig. 9 shows Vern Glass' spar completed and the author's ready for the closing up.

Fig. 9

In summing up, we are happy to say that we encountered no particular difficulties and both spars turned out to be accurate. Vern's spar was slightly heavier than mine only because it was intentionally made several millimeters wider; the weights were 52.5 and 55.5 lbs. The total amount of labor **per spar**, including jig construction, turned out to be 200 hours, which does not include time spent beforehand by Vern in building and leveling the work bench.

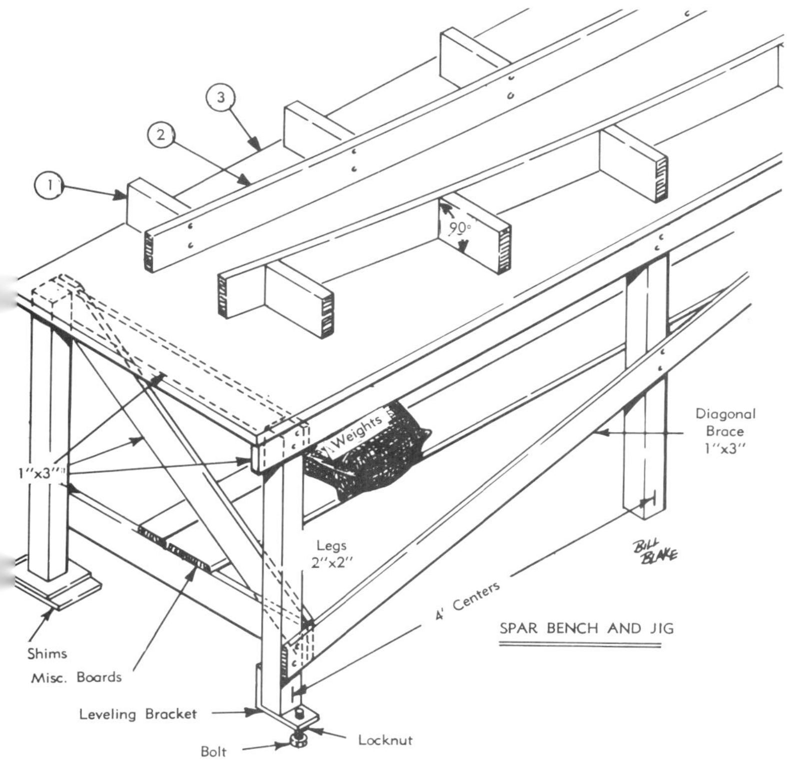

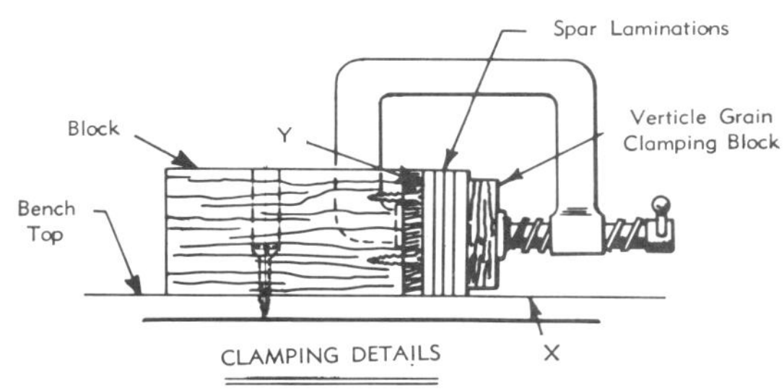

Fig. 10

Excellent workmanship is evident on this all wood low wing aircraft as it nears completion. Note built up main front spar. A hollow spar can provide tremendous strength with minimum weight. Many aircraft, particularly those of European design, are built in this manner. Extensive jigging is required to get the built-in dihedral and wing tip wash out. Another factor for builders to consider is available work space as these spars are usually built in one piece and therefore require considerable work shop area.

Gluing Can Be Easy

By Clint Lillibridge, M.D., EAA 18847

HAVE YOU EVER SAID, "Nuts! It's past upper limit of closed assembly time, and I'm not finished clamping." Or, "What do you mean, 'The pot of glue is all set up'?" Many flying enthusiasts are not homebuilders primarily because of the ritual required by glues currently in widespread use in this country. Our own EAA manuals are replete with the rites of ambient temperature, the witchcraft of sanding vs. planing of surfaces to be glued, meticulous notes regarding open times vs. closed times, and exhortations on clamping pressures.

Many homebuilders here in Chapter 26 have followed the lead of hundreds of our Canadian colleagues by changing to Aerolite Glue (made by Ciba Ltd. of Canada). They are now building wooden aircraft which they probably would not have attempted if the casein glues or resorcinol-formaldehyde adhesives were the only materials they could use. As Hugh Beckham noted in his article "Fifinella Finally Flies", ** Aerolite glue makes the use of other glues seem both cumbersome and capricious in comparison.

Regular users of Aerolite concur with the manufacturer's statements regarding advantages:

(1) Joints having gaps up to 1/16 in. are of sufficient strength that the wood ruptures before the glue line separates.

(2) Clamping pressure of a pound or two is quite adequate. Staples from a hand tacker join pieces sufficiently to insure complete setting and bonding.

(3) Open assembly time may be extended almost indefinitely, and is relatively unaffected by temperature.

(4) Closed assembly time is unrelated to length of time used in open assembly.

(5) Moisture content of wood does not alter final glue strength (although increased moisture prolongs setting time).

(6) Gluing may be done safely at temperatures as low as 50 deg. F.

(7) The final product is entirely waterproof, and withstands open exposure for many years without deterioration.

(8) One may choose among three different accelerators to get different rates of setting (5, 10, or 20 min.) and curing (½, 2, or 5 hrs.).

As with all products, there are some disadvantages:

(1) Poor adhesiveness to concrete, metal, or plastic.

(2) Joints cannot be boiled. (But who boils airplanes?)

(3) Thought must be given as to which side of a joint one will apply hardener and to which side the resin, to avoid incompatibilities later.

Perhaps we should clarify some of the information which was circulated at Rockford this past August, by well-meaning sources who may not have had extensive knowledge of Aerolite. Basically, Aerolite is an unusual form of urea-formaldehyde adhesive. This form of glue usually has two major defects: hot presses are required, and gap joints deteriorate very quickly. Both stem from the uneven manner in which the resin polymerizes. Early solutions to the problem involved compounding fillers in the resin. These delayed, but did not prevent, glue line failure. The Aerolite system, on the other hand, works on an entirely different principle. The accelerator is placed on one side of the joint, and the basic resin is applied to the opposite side. No reaction occurs until the two pieces come into contact. This process gives controlled polymerization, and obviates the two primary defects of urea-formaldehyde glues.

The joint may be left open as long as the water-soluble components don't evaporate to dryness. A fresh brushful prolongs open assembly time almost indefinitely. Since glue components never meet until the joint is assembled, the time required for setting depends only on the accelerator used, and the temperature.

The basic resin is stable in powder form for years, if kept dry. The resin when mixed with water is stable for many months if kept cool and tightly covered. If reopened frequently in a warm room, the resin is usually good for at least a few weeks. The accelerator is good indefinitely.

"Does it work?" the dubious homebuilder asks. I had very serious doubts about it the first time someone mentioned the technique. After using both the phenol-formaldehyde and the resorcinol-formaldehyde glues with excellent results for many years, I thought that the claims made by the Ciba people must be unfounded. For the past several months, I have been performing a set of tests to demonstrate the limitations of this "impossible" gluing system. Test specimens of ½ in. X ½ in. spruce were mated where the only gluing pressure was the weight of the wood itself. Other specimens were soaked for one month under water after gluing. Still other wood was soaked before gluing. Some wood surfaces were left as rough as they came from the saw, while others were planed and still others were sanded. After applying glue, some were left unassembled all day (during which time the hardener dried completely, but the resin remained tacky). These were recoated in the evening, and then assembled. Some joints were left with 1/16 in. between the two pieces of wood. In some specimens, end-grain was butted to side-grain.

In spite of these many attempts to get poor joints, we found that the wood ruptured a short distance away from the glue line, which remained intact. A few speci-

** *SPORT AVIATION 1965*

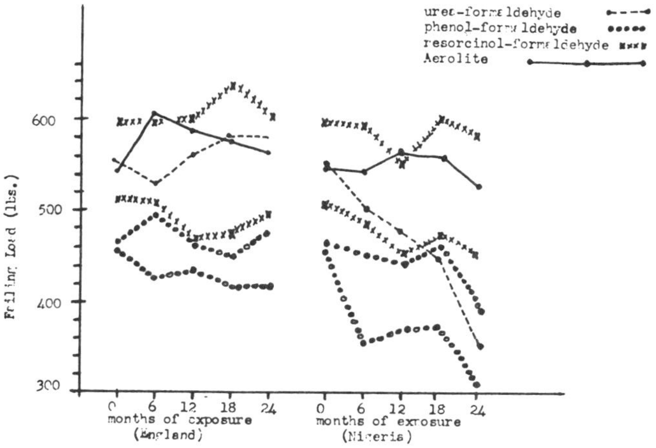

Each point is the average of 12 tests, recorded in lbs.

FIG. 1

mens in which the clamping pressure was so light that the accelerator and the resin did not make good contact over the entire glue area resulted in separation at the glue line. Much to my amazement, the advertising claims were correct!

The Rutherglen group of the Popular Flying Association in England has made similar destructive tests, and found about the same thing. (Results of their testing is available in Aircraft Research Laboratories, Technical Notes No. 183.) Aerolite complies with British Standard No. 1204, and is fully accepted by the Canadian Department of Transport. Several aircraft built in the USA using Aerolite have received FAA approval.

A specific aging test was set up by the Ciba Co. in England. Data of these tests was obtained from ARL Technical Notes No. 133.) Two identical sets of specimens were made using several different types of aircraft glues. One set was sent to Port Harcourt, Nigeria, and the control set was kept in England. Specimens were kept in special boxes which simulate aircraft wing interior conditions. These boxes were mounted three feet above the grass, fully exposed to the weather. Temperatures within the box in Nigeria ranged from 69 deg. - 106 deg. F. Humidity measurements during these tests ranged from 51 - 95 percent. Moisture content of the wood increased from 11 percent to 18 percent. Samples from both sets were tested every six months in England. The results of the gap specimens showed the most significant differences, and are given in the graph. Each point is the average of 12 tests. The standard American adhesives withstood this weathering poorly. The resorcinol - formaldehyde which exceeded Aerolite is a new product from England.

In summary, Aerolite is the kind of glue which enables the first time homebuilder to take on a difficult project successfully, and helps the experienced builder complete his aircraft with dispatch. Homebuilders no longer need be hampered by the usual cumbersome gluing techniques. If you don't believe it, try some yourself! As a result of doing these tests, I have given up using my old favorite resorcinol-formaldehyde glues altogether.

Popular with homebuilders are scaled down replicas of fighter planes from World War II. The Corsair F4U shown here is just one example of many of these types that show up at EAA gatherings. These replicas are designed so that their basic construction is wood which provides the required strength for carrying flight loads. Over this basic framework of wood foam or other composite material is added to form desired fighter plane contours.

Tips on Aircraft Woodwork

By Arthur W. J. G. Ord-Hume

In aircraft construction it is often required to bend wood to a curve. There are two methods of forming wood to a bow - one is by bending and the other is by laminating several thin strips of wood together in a jig.

Both spruce and ash bend readily to a certain degree. Any bend in excess of this amount will result in fracture of the fibres. Also, when bent, the wood is sufficiently elastic to cause it to spring straight again. By softening the wood, however, not only will the timber remain bent but it will be possible to increase considerably the safe radius of the bend.

Wood may be softened in two ways - water softening, which may be done either hot or cold, and by steaming. The latter is the most effective but, since it usually necessitates the use of a steam-box if the piece is long, it is more usual to plasticize the wood by saturation in water.

Soak the portion to be bent in water for up to two hours until it is soft enough. Bend it very slowly to the shape required, working the bend in with the fingers to prevent sudden stretch of the fibres.

The process is speeded up by immersing the piece in boiling water or, if the piece is small enough, actually boiling it in a pan of water for a few minutes. Remove the wood with gloves and again bend it slowly. If the bend is acute or the material thick, repeat the boiling treatment until the bend is completed.

It will be found that upon drying there is a tendency for the wood to spring slightly out of the bend. To counteract this, increase the bend from 15 to 20 percent more than is required so that the wood when dry will adopt a "set" just about at the bend required.

In steaming wood to be bent, the effect on the wood is similar to that when the wood is treated as above. The fibres soften readily without any loss of strength upon drying. If only a small area of wood of thin section has to be treated, it can be done in the steam from the spout of a kettle. However, for larger pieces a steam-box must be resorted to. This consists of a long metal box or large-diameter tube of sufficient size to accommodate the wood to be steamed. It is sealed at one end with a sealable door at the other. The box or tube is connected via a pipe to a water boiler heated by a gas burner. As the water boils, the steam is forced under pressure into the box containing the timber. A suitable safety valve must be fitted into the boiler. Great care must be exercised when exhausting the steam from the box - a scald from steam is a most serious and painful experience.

Timber in a steam box should be left from one to two hours before removing. Subjection to longer periods of steaming results in the destruction of the suppleness of the wood after it is dried. Before glueing any wood which has been steamed or wetted in any way, always let the piece dry thoroughly in the open air. On no account try to accelerate the drying by artificial methods. Air drying will preserve the correct moisture content of the timber.

Plywood may be bent in the same manner. It is easiest to bend ply with boiling water saturation. Bend the ply as far as it will safely go and, while still holding the bend, get an assistant to pour boiling water from a kettle over the outside of the bend. Immediately the ply will soften and start to fold tighter. Do not hurry the process, for the wood may yet split. By repeated applications of boiling water, gradually work the ply to the curve desired and then clamp it between two boards to dry overnight.

Where it is desired to make a large curve of accurate dimensions with a comparatively thick cross section, the designer will usually call for the piece to be laminated. Formers, wing tip bows, keel members and similar pieces call for lamination.

A jig must be made to locate the strips for glueing. Obtain a large flat board of commercial wood sufficiently large to take the curve of the member to be made, plus a margin of several inches. For very large members, a wooden floor may be used. With a hard pencil mark out the curve. Now cut wooden blocks about 1 in. by 2 in. by 3 in. and glue and screw these to the inner edge of the curve at intervals of every 3 or 4 in. If the curve is sharp, these blocks will need shaping. To prevent the adhesion of the glue to the jig, paint the jig with two coats of hot paraffin and allow to dry.

Cut the strips of wood to be laminated and allow several inches more in length than is actually required - the excess may be removed later. Glue up the laminations separately on a flat surface, shuffling each strip to the next to exclude air. When the desired number of pieces has been glued up, place the bundle into position on the jig board. On the outside of the bundle, opposite each block, put a piece of scrap wood to use as a clamping block. Alternatively, a strip of steel may be bent into place on the outside rather like an additional lamination to take the pressure of the clamps.

Clamp up each block with C-clamps and leave it for 24 hours to dry. After removal and cleaning up, examine for any signs of de-lamination. If there is any gap, run thin glue into the split, working it well in with a thin strip of metal such as a feeler gauge. Reclamp and leave until dry.

When sawing aircraft timber, support it adequately to prevent it from splitting when the saw approaches the other side. Where possible, clamp a piece of scrap wood to the underside so that the saw finishes its cut in that timber rather than in the required piece of wood. This produces a clean cut. Similarly, when cutting plywood, the saw will leave a rough, split lower edge. It is not always possible to "back up" a cut in plywood as detailed above but by cutting no closer than 3/32 in. to the line, this rough edge may afterwards be removed with plane and sandpaper.

Remember that wood splits easily and is weakest along the grain when it is being cut. If one is attempting to cut along the grain or at a slight angle to it, take great care for if the saw slips or one tries to hurry the cut, the wood will split, invariably ruining the portion wanted.

Planing wood is another operation necessitating care. Always plane with and not into the grain when surfacing a piece of wood. Examination of the grain will show that, even on the best boards, it tends to converge slightly to one edge. If one tries to plane toward the grain, the blade will dig into the end of the grain where it meets the surface. Run the plane so that it slices the grain from behind.

Should it become necessary to plane the end of a plank or piece of wood so that the plane is cutting at right angles to the grain, the blade must be very sharp and set so as to remove the smallest thickness possible, otherwise it will dig in. The forward edge of the piece of timber should be backed up with a piece of scrap wood so that the plane continues its slice into this. This is to prevent the edge from being split by the plane blade.

The plane may be used for the shaping of a curved member if it is an outside curve. However, it is more usual to make use of the spokeshave with which one may also plane an inside curve. It should be held firmly in both hands and drawn towards the operator. Templates may be made out of plywood to check the shape of the work.

To smooth wood or to remove glue that has dried, do not use a joiners rasp or dreadnaught. These will tear the grain. A coarse flat file is the best thing to use, although it will clog quickly. The file should be used to smooth off wing tips and so forth in conjunction with the sandpaper block. Clean it frequently with a file scratch card. Should the file become clogged with glue, immerse the file in boiling water for a few minutes and then clean in the normal way.

Store aircraft timber on a rack or shelf away from drafts and dampness or heat. Never keep wood on a concrete floor. Avoid contamination with oil, dirt or grease during storage and work. Woodwork is clean work. Don't work with dirty hands. If it is necessary to store incomplete or uncovered components for awhile, protect them from dust and dirt by using dust sheets.

The essence of good aircraft carpentry is careful workmanship, the accurate following of drawings and, until experience is gained, an unhurried approach. The best instructor in the world cannot supply the most vital ingredient - experience. Both carpentry and metalwork are subjects which demand practice and experience to gain proficiency. That goal may be reached with patience and care.

Remember - sound work is a sure step towards success.

Building a box spar involves lots of room to work — plus accurate jigging. There are many types of construction for hollow main spars — the example shown above clearly illustrates the many individual pieces that are required to carry the loads for this particular aircraft. When all parts are glued inside the spar a laminated plywood sheet will be added to fully enclose the box spar.

Laminating a single piece spar requires a lot of clamps. This builder has made his own clamps out of sections of 2 x 4s and lots of long bolts.

A Dual Purpose Sander

by Richard J. Halstead

It wasn't long after starting an all wood airplane (Taylor Monoplane) that I discovered the need of a means to do precision sanding in the fitting of dozens of braces and corner blocks. The need for a precision scarfing method reared its ugly head later after the fuselage was all framed.

In an attempt to keep tool cost down and conserve my meager airplane "Kitty", I decided to build a sander instead of investing in a commercial one. Performing the proverbial search through my pack-rat inventory, I came up with most of the material shown in the drawing. The only things I lacked were the sanding heads and mandrels. A spare ⅓ hp capacitor start motor was fitted to a wood stand made with 2" x 2" fir legs banded with ¾" x 3½" pine screwed together. The sanding table and supports are also ¾" pine stock.

To get the proper operation of the table, a precision means of adjusting its angular position to the sanding disc was accomplished by utilizing a standard crank out window assembly, an item available at any large hardware store. An adjustable tension lid support locks the table in any given position. The table is mounted with the hinge centerline passing across the disc face and the table just clearing the top of the mandrel nut. A metal sanding disc was found in Sears tool catalog and its performance was much superior to a plywood disc. The disc was mated to the ⅝ motor shaft with a brass mandrel bored ⅝ dia. on one end and turned for ½-20 THD on the other end. I couldn't find a stock item for this part, although I have seen them on occasion. The aluminum oxide paper on the disc is contact cemented and replacements are available.

This assembly put precision in my glue joints and really built up my neophyte confidence. My confidence was quickly shattered, however, on my first try at scarfing a six foot length of plywood for the fuselage skin. I attempted to scarf it on my table saw with the sanding disc and a sliding vertical fixture. After watching the plywood slowly disappear down in the table between the disc and the table insert, I knew I needed an easier method.

Scarfing plywood in a horizontal position seemed a safer approach so back to the sander for a modification. The Sears catalog yielded a 2½ dia. x 3" lg. sanding drum with replaceable sleeves. This was mounted on a second identical mandrel (a common one could be used) and interchanged with the sanding disc. The table was obviously now in the wrong position so a second set of mounting holes (Labeled 2) was drilled in the table supports to allow the back edge of the table to just "kiss" the bottom of the drum, while the front edge of the table was depressed about 5° in its full down position. This method allows the table to be slowly cranked to a perfect 16 to 1 scarf slope. (approximately 3.5°) The addition of a scarfing fence maintains a constant scarf.

The safest procedure for scarfing was found to be: (1) Always use a sample piece when setting up to insure proper adjustment. (2) Feed the sheet back into the drum (against direction of rotation) in progressive passes. Don't try to take all the material off on the first pass. This will insure the smoothest consistent results. This is really important when scarfing parallel to the outer grain. It gets very fragile! (3) If large sheets are being worked, think about using the longer optional table.

The conversion from sanding to scarfing involves pulling the four bolts in the table supports to reposition the table and also changing from disc to drum which amounts to a five or ten minute job. The sanding sleeves on the drum have unusually long life when used for scarfing and only two or three of medium grit were needed for the whole skining operation on the monoplane which is almost completely covered with ply.

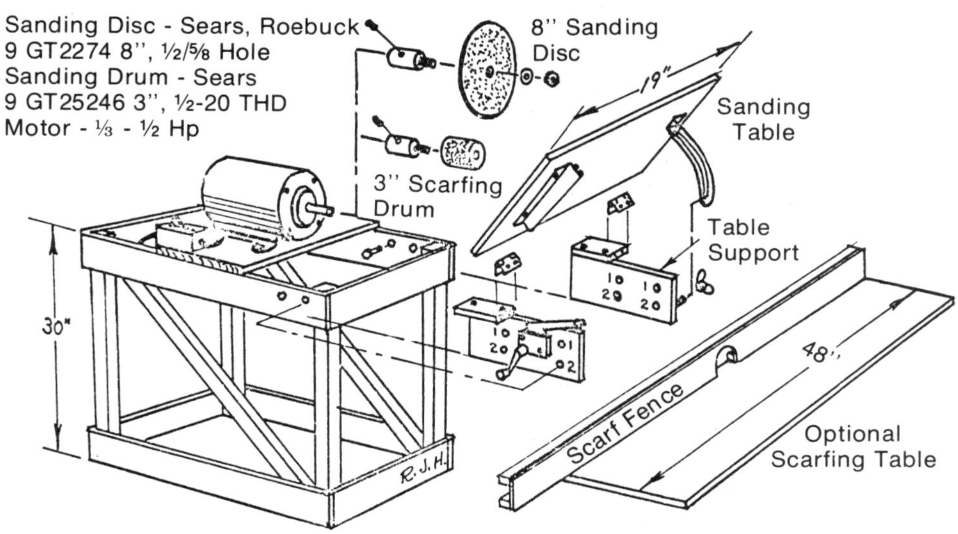

First of a series of articles on

Aircraft Woodwork

by Arthur W. J. G. Ord-Hume

Aircraft carpentry is by no means difficult to master - it is no more exacting than the making of good furniture or a bookcase. While an article of furniture depends not only on sound workmanship but a good superficial finish, the wooden aircraft is more dependent upon careful construction. If the work is executed properly, then the finish, though achieved by entirely different methods, can be good.

The aircraft carpenter must be capable of manufacturing a part or an assembly to close limits out of wood, plywood or a combination of both materials. Before dealing with the actual carpentry, however, we should consider two dumb assistants who each play a vital part in aircraft woodworking. They are therefore worthy of detail. One is the brad (known variously as gimp pin, tack or nail) and the other is the wood screw.

The aircraft brad is a sharp-pointed nail with a flat head. It is usually made of brass, but can be made of cement-coated ferrous metal. It is available in lengths from ¼ in. to 1 in. and in thickness from 22 s.w.g. to 18 s.w.g. It is different from the commercial nail in that its flat head does not cut the surface of the timber when it is driven home. This flat head also increases the pressure which the nail can exert on the timber.

As a general rule, brass brads are used only to secure a joint until it is dry. When attaching plywood to a structure, it is common practice to tack it on with brads and remove them when the glue is dry. With modern synthetic resin glues, no great advantage is to be gained by leaving them in place, although they will not corrode or damage the wood by their presence if they are left in place.

A word of warning - when using brads, *never carry them in the mouth!* They are indigestible if swallowed!

In place of brads, ordinary staples may be used where they can definitely be removed after the glue has set. For the use of staples, a good office stapling machine should be used, and the base either folded back or removed altogether by the withdrawal of the securing pin at the joint. The machine is then held firmly in contact with the wood and the staple driven home by pressing the knob with the other hand.

In the removal of brads or staples, exercise care so as not to damage the fibres of the wood in the process. A useful tool for this purpose can be made out of an old screwdriver. Soften the end by heating, file the blade to a sharp edge, cut a cleft in it (similar to a claw hammer) and bend the end through 60°. Retemper before using.

The wood screw is used for the additional attachment of certain parts, such as wingtip bows, inspection panels, small hinges, seat members and so forth. Screws for aircraft use are made of brass or cadmium-coated steel. They are available in round-head, countersunk or pan-head style and should be used as specified on the aircraft drawings.

Wood screws are designed to cut their own threads in the timber and derive their holding power from these threads. If the screw is driven into the wood without a pilot hole, the wood will probably split. If the pilot hole is too big, the threads cut by the screw will be insufficient to hold the screw tightly.

Pilot holes for screws should be drilled to enable the screw to be used to its maximum advantage. Two drills are necessary - one to drill the hole for the shank and the other, smaller in diameter, for the threads.

The first hole should be very slightly smaller than the diameter of the screw shank. If it is too large, the shank will not be a tight fit and the entire load will be thrown upon the head of the screw, which may pull through the timber. If the clearance hole is too small, the shank will bind so tightly that the threads will be unable to draw the screw into the hole and they will ream out the second hole.

This second hole must be of the correct size to allow the threads to get a proper grip in the wood. If it is too large, all the wood necessary for the screw to bite into will have been removed and it will not hold. There again, if the hole is too small, the wood is likely to split. Also of importance is the length of this second hole. If it is too short, the screw may tend to hold the members apart or fail to draw them tightly together. This is particularly the case with hard, high-density woods.

Carefully measure the screw to be inserted and select two drills accordingly. If many screws of the same size are to be used, fit a stop onto the drill shank so that all the holes will be of the same depth. A piece of wood such as a dowel rod will do for this. If the screw to be used is a countersunk one, use a proper wood screw countersinking bit. Woodscrew heads have an angle of 90°. Never use an ordinary large twist drill as a countersink. The head angle of the drill is 120° and the edges of the drill will chip the surface grain of the wood.

The drilling of a hole in wood is a particularly important operation demanding the utmost care and accuracy. Careless drilling can produce a weak and often dangerous structure. Each fibre in a wood strip has a job to do and the designer has calculated just how much of the wood may be removed by drilling a hole without affecting the strength of the structure. If the hole is made oversize, not only has too much wood been removed but the bolt or fitting which passes through the hole will not be supported evenly by the timber. A bolt in particular depends upon the bearing surface of the wood through which it passes to enable it to resist the loads upon it.

Basically, to drill the wood the drill must be held rock-steady at the precise angle to the wood which is required. If it is attempted to drill a hole half way from each side, the chances are that the two holes will not meet dead in the center. If the hole is reamed out to fit a bolt, the hole will be barrel-shaped. The bolt would only be supported at its extremities and, when any load was applied during service, the bolt would very rapidly elongate the hole and work loose.

The correct method is to drill the hole entirely from

one side. If the drill is slightly crooked, the hole will not mate up with the fitting on the other side. Any attempt to rectify the matter by reaming or redrilling from the other side will be useless, since the lack of bearing surface for the bolt will allow the fitting to be pulled out of place. Should there be any doubt whatsoever as to the accurate placing of a hole, drill a small pilot hole first to check with.

A hole is an extremely deceptive thing. While its outward appearance may seem satisfactory, it is often impossible (except with large holes) to inspect its inner surface to verify whether or not it has been properly drilled. It rests largely with the constructor to exercise caution and patience over this job.

For general purpose work, a satisfactory drilling jig may be made by boring a block of steel or iron about 2 in. by 3 in. by 2 in. on a proper pedestal drilling machine with a hole to the required size. It is a sound scheme to drill the blocks with about four holes, say 3/16 in., ¼ in., 5/16 in. and ⅜ in. in diameter, quite closely spaced near the center, thereby enabling the jig to be used for a variety of holes. The local garage will in all probability drill the block on a proper machine for a few cents.

In use, first draw the center lines locating the hole on the wood, then place the jig over this mark, sight down through the block to double check on the positioning and then *lightly* clamp it into place with C-clamps to prevent it from moving.

It is recommended that drilling should be done with an electric drill, but if a hand drill must be used, obtain one with a breast plate as it is easier to hold steady. Drill smoothly and avoid "waggling" the drill in the hole. Always rest the work on a solid block of wood so that the drill passes through into it after boring the job. This is to prevent the splitting of the surface grain on the other side. Alternately clamp or press the block by hand to the other side, taking care to keep the fingers clear of the drill tip should it break through the other side.

Sometimes it is not possible to drill through a jig as detailed above. In such cases, the help of one, preferably two, assistants will be needed. Obtain two squares (machinist or carpenter, about 6 to 8 in. high is ideal) and stand these up near the hole to be drilled, at right angles to each other from the hole. Hold the drill approximately vertical with the point on the center spot of the hole. The assistants may now sight whether the drill is truly vertical by sighting against the vertical arm of each square. When both are satisfied that the alignment is correct, proceed slowly to drill the hole, pausing frequently to recheck with the assistants.

Varnishing With the Pump

by Christopher McGuire

Whether rebuilding an antique airplane or building one from scratch, anything we can do to maximize our efforts will make the process go more quickly and be more pleasurable.

Varnishing the myriad pieces on wood structured aircraft can be a laborious task. The many gussets, braces, ribs, spars, stringers, etc., consume a lot of time. Much of the time is wasted to and from the can. On one of my projects, I decided there must be a better way and thought of this idea. It is a great labor saver and only takes a few minutes to make.

First, I drilled a hole in a small brush. The hole is angled to the root of the bristles and sized so that a 1/8" I.D. x 3/16" O.D. x 1/32" wall clear vinyl tube can be epoxied in the brush. Next, I cleaned my pump oil can and attached the other end of the tubing to its spout.

To use, simply fill the can with varnish, pump with one hand and brush with the other. The technique is to pump just enough to keep the brush at the optimal level of saturation for even distribution. Of course, this device could be used to apply many other coatings as well.

Some advantages are:

1. 2-3 times faster than conventional methods.
2. No overspray if one were to attempt spraying.
3. No dripping while transferring brush to and from can.
4. No lost time dipping brush.
5. Never lose eye contact with work, therefore, less chance to miss a spot.

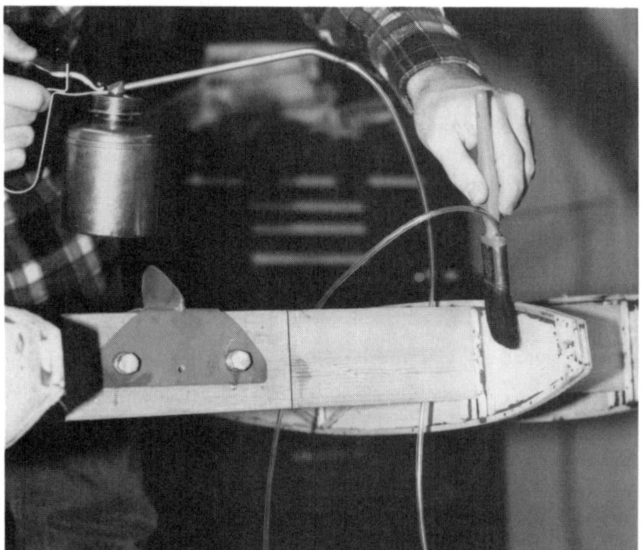

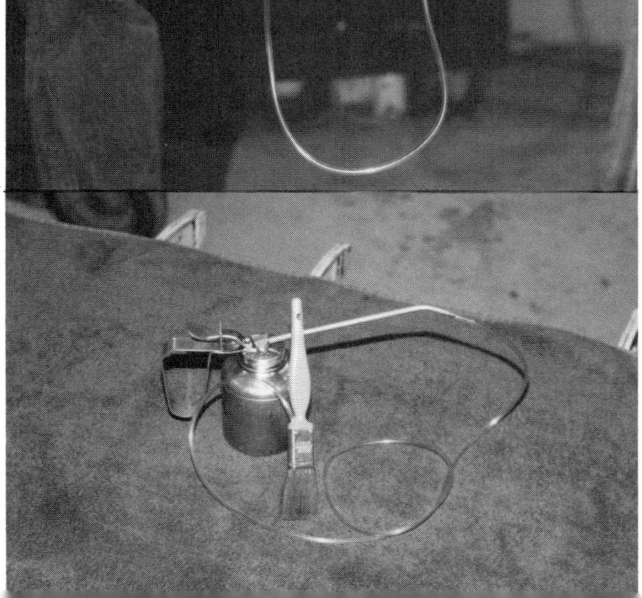

Joints for Aircraft Woodwork

By Arthur W. J. G. Ord-Hume

In aircraft carpentry there are only two basic types of joints used. Each of these depends on careful workmanship to enable it to develop its maximum strength.

First is the scarf joint, in which two pieces of wood are equally tapered over a precise length so that they fit together, one onto the other. The scarf joint, properly made, will at least equal the strength of the surrounding timber since it presents a very large surface for glueing. For example, a one in. sq. spruce strip joined by a scarf joint with a one in 16 taper has a glue area of 16 sq. in.

The scarf method of joining is the only method which may be used for joining or splicing spars, longerons, stringers or any other such strip of timber. It is also used for the joining together of two panels of plywood to make one.

The width of a scarf is normally 16 times the thickness of the wood - a piece of 1/16 in. thick plywood will need a scarf one in. wide (16 x 1/16), and a strip of ¾ x ¾ in. wood will require a 12 in. scarf (16 x ¾).

When joining a piece of timber of unequal section such as ½ x ⅞ in., one can either scarf the ½ in. face 8 in. long or the ⅞ in. face 14 in. long. For reasons of economy, however, it is usual to make the scarf only as long as necessary. In all cases timber which has one face three or more times wider than the other (for example, a plank 4 in. x 1 in.) must be scarfed on the broadest face. The above example would have a 4 in. wide scarf 16 in. long.

The accurate cutting of a scarf requires a really sharp smoothing plane with a very fine setting. After marking out for the scarf using a carpenter's square, rule and pencil, the timber may be sawed to the approximate angle with a tenon saw. Care must be exercised to avoid splitting the wood during this operation.

Secure the required portion of the timber to the bench with the longest edge of the scarf on the bottom and supported by a scrap piece of wood with a straight edge. This is so that as the thin edge is prepared, the end is supported and remains level with the lower edge of the timber. If it is not supported, it will bend down under the weight of the plane and the scarf will not be true. Firmly locate the timber and the supporting board to the bench with C clamps to prevent its moving.

The cutting of the scarf will be facilitated if the plane is held at a slight angle to the direction of the cut so that the wood is sheared off. This angle may be about 15 to 20 deg. When scarfing plywood, it is important to keep the cutting edge at a convergent angle to the cut or else the blade will dig in and chip away this very fine edge. The plane should be set to give the finest possible cut. A coarse setting will chip the wood and fail to produce good work.

Take great care while planing to ensure that the surface being produced is level and true. Use a square and rule and check frequently. When both pieces have been prepared, try them together as a final check for accuracy.

In scarfing plywood, the finished edge should show the laminations as straight bands. The glue used in plywood manufacture is a synthetic resin which will show up as parallel dark lines. If these glue lines are not parallel but are converging, wavy or kinked, then the scarf is not true. This can often be caused by faulty planing or, as usually is the case with kinks, by an uneven baseboard, dirt on the underside of the plywood sheet, or by the presence of knots in the baseboard causing the cutting edge of the plane to lift at one point.

When glueing a scarf joint, one must apply even pressure along the entire length and width of the joint.

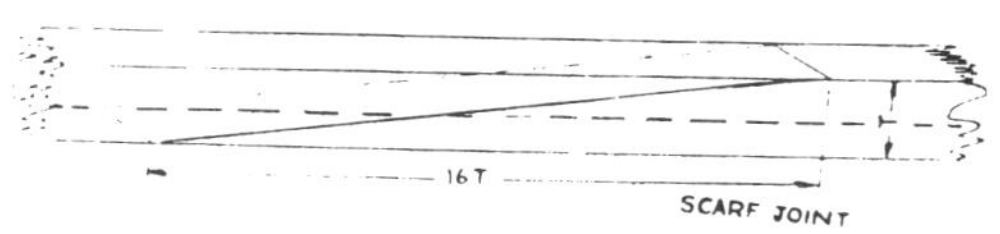

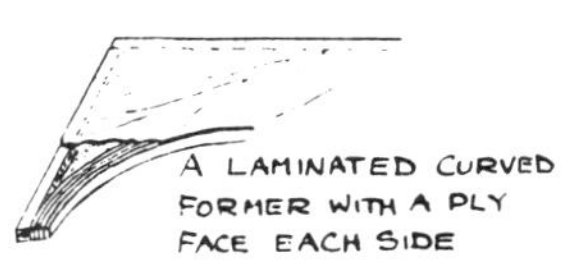

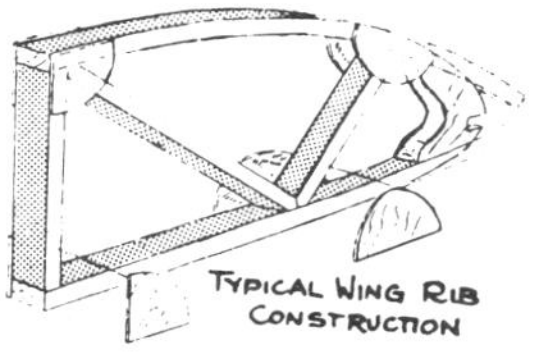

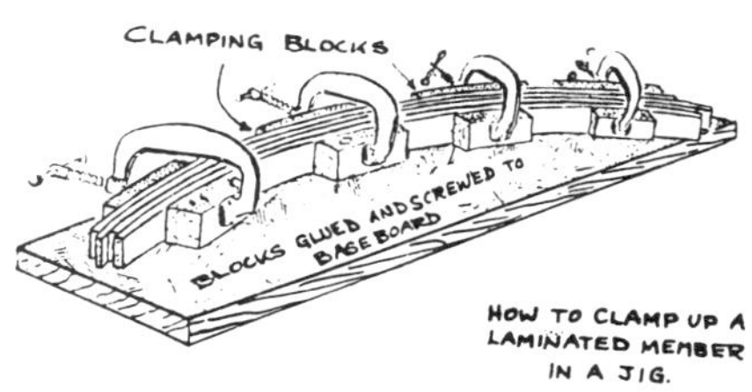

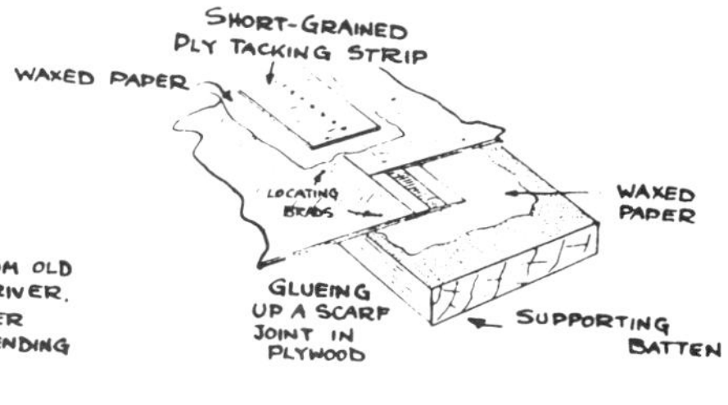

TYPICAL JOINTS IN AIRCRAFT WOODWORK

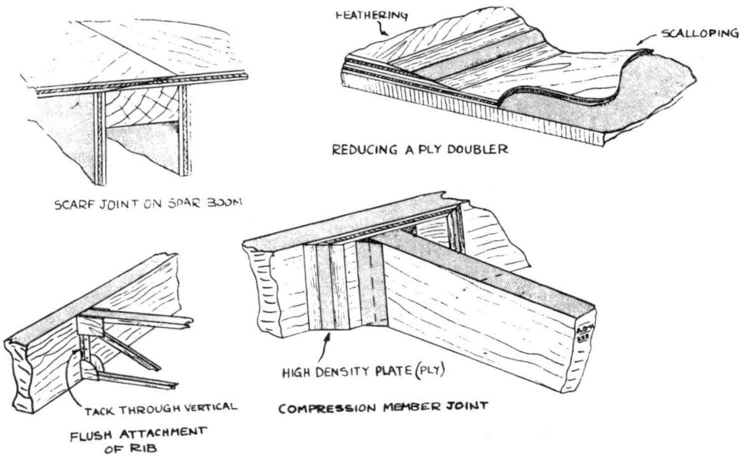

The glue should be applied evenly to both faces and then the two pieces rubbed together to exclude any air bubbles. Select two pieces of scrap-wood several inches longer than the total length of the joint. Lay a sheet of grease proof paper over one strip and place one piece, scarf uppermost, onto this. Apply the glue and mate the two surfaces together. Carefully make sure that the two pieces are in the correct position. Place a piece of grease-proof paper over one strip with the second piece of scrap wood upon that, clamp up the joint with C clamps. Clamp at close intervals to ensure adequate pressure. Achieve *adequate,* not *excessive,* pressure. Never put a clamp directly onto aircraft wood - it will press into and damage the wood. The load should always be spread by transmitting it through clamping blocks.

Joining plywood presents a slightly different problem. Due to the size of the panels of plywood which may have to be joined, it is nearly always impossible to clamp the two pieces together as detailed above. Let us suppose that the constructor has to scarf up two panels of plywood and the scarfed edge is to be 20 in. long. First of all a batten of wood will be required about 3 in. wide, ¾ in. thick and 2 ft. long. A strip of grease-proof or waxed paper is laid over this batten and one panel of plywood laid with its scarfed edge uppermost and approximately on the center-line of the batten. Tack the body of the plywood every 6 in. or so to the edge of the batten. Apply glue to the joint, shuffle the other panel onto the scarf to exclude air and line it up in the correct position. Secure it to the batten with tacks as you did on the other piece. Place a strip of waxed paper over the joint and prepare a strip of plywood as long as the scarf joint and about 1¼ in. wide. This must be short-grained so that it will press the full width of the joint. Line up this strip and proceed to tack it at half inch intervals along its center-line (the center-line of the scarf joint). Use brass gimp pins of sufficient length to go right through the joint and into the base batten.

When the glue has set, the tacking strip may be removed and the odd securing brads removed. If properly made, the joint should require only a light sanding to remove any surplus glue.

We have now learned to use one of the most versatile applications of light clamping, the tacking strip. It is a method of applying sufficient pressure to set a glued joint when it is either impractical or impossible to use a clamp or other means of pressure application. When covering the leading edge of a wing, for example, it is impossible to use clamps for securing the plywood since both sides are not accessible. In such cases the tacking strip has to be used.

Strips may be prepared from off-cuts of 1/16 in., 3/32 in. or ⅛ in. ply and should always be short-grained unless they are for special use. The whole idea of the tacking strip is that the fibre of the major (outer) veneers should press down on the joint or center of pressure application. This means that the grain should be at right angles to the line of pressure wanted. Do not forget that any glue squeezed out of a joint will stick to the tacking strip or clamping block. Use waxed paper wherever there is a risk of this happening.

The second type of joint in aircraft carpentry is the butt joint, several combinations or types of which are used in aircraft construction. A butt joint itself is not strong. It presents the smallest possible glue surface. However, it is often the only method of joining a strut to a longeron or a rib vertical member to a capstrip. The butt joint alone, therefore, is of little strength and other means must be resorted to, to enable it to develop strength. These may be corner blocks or plywood gussets which will increase the glue area and provide positive results from what is inherently a weak joint on its own.

The butt joint is most commonly used by the designer to resist only compressive loads. In such cases there is little or no additional force

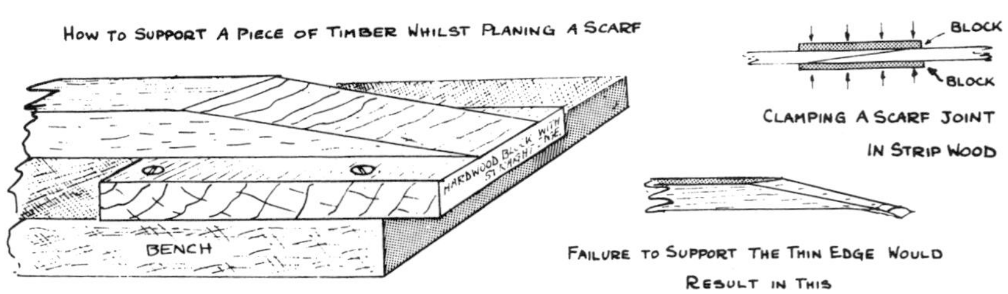

PREPARATION OF A SCARF

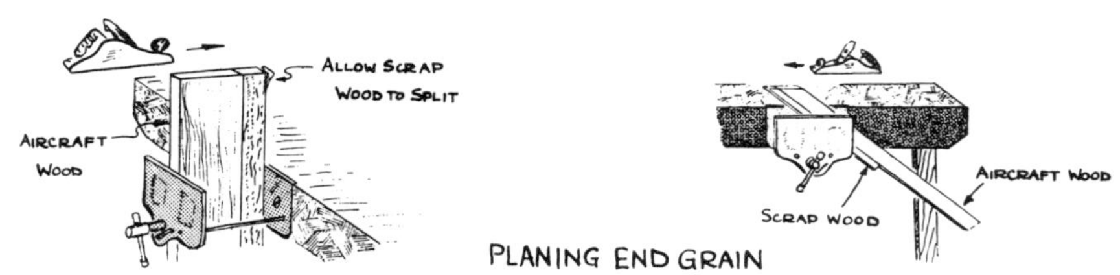

PLANING END GRAIN

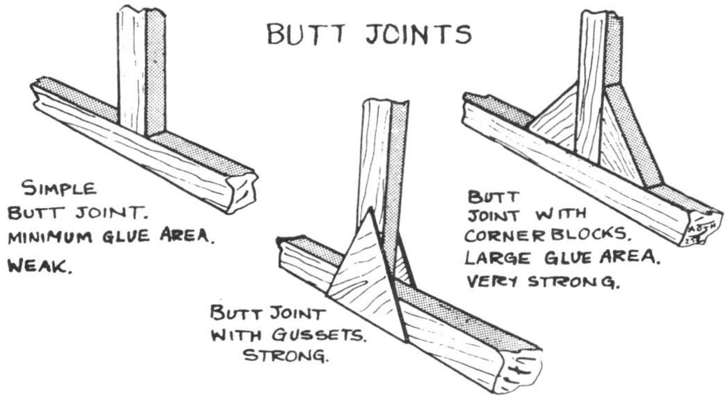

at work on the joint likely to cause it to be displaced and it will suffice to hold the member in place using small plywood gussets.

Where the plywood skin of a structure is employed to withstand shear loads, the cross member once again has only to resist compression. However, in this case it is attached along its length to the plywood skin. This not only serves to stiffen the structure but also reduces the area of unsupported thin plywood which would otherwise bulge. The designer evolves the minimum size of an unsupported panel which he can utilize and he will assume attachment to the cross member.

When other forces are at work on a joint, the designer will make use of solid corner packing blocks to increase the cross-section of the members at that point for glueing. He may also use gussets or a combination of corner blocks and gussets. In some cases he may resort to a metal fitting, bolted to the joint, to take the load. In this latter case it is assumed that the joint is "dry", or unglued, although in practice it will be glued.

Where a member has to take a direct tension load which has to be absorbed into a transverse member (for example, a wing pylon attachment to a top longeron) it is uneconomical and impractical to develop strength by gussets since they would need to be excessively large. In this case the member is attached by a metal fitting. To reduce the possibility of the bolts tearing out of the member under load due to the shearing of the end grain, the bolts may be staggered so that each bolt is supported by the maximum possible unbroken length of grain. An additional or alternative method of stabilizing the end grain against shear is to face the member on both sides with plywood, thus allowing the bolts to pass through the ply as well as the actual member.

Designer Paul Poberezny stands beside the nearly completed left wing panel of the Pober Pixie. Trailing edge aluminum is clamped into position ready for nailing. Right panel behind Paul is in the initial stages of assembly.

Part 1

Jigs for Wooden Aircraft

Fuselages

By Arthur W. J. G. Ord-Hume, EAA 8579

Arthur W. J. G. Ord-Hume with Luton Minor (pre-war model) which he built in 1949. The ship is still going strong. The above photo was taken at Panshanger Airfield near Hertford last summer.

Before constructing any form of jig, the first consideration is just exactly what is the jig for. What is it intended to do and how much can be expected of the ideal jig.

Broadly speaking, the jig is the airplane constructor's 'third hand', a hand which is capable of positively holding with extreme pre-determined accuracy a number of parts all at once and in the correct juxtaposition. It is your 'third hand' because it will faithfully hold tubes for welding in the right position, locate parts for skinning and keep the sides of the ship true for marrying up.

A jig can be made from almost anything from fiberglas through to chipboard, wire, bricks and even concrete, papier-mache and plaster. The floor under your feet can be turned into an almost astronomical number of jigs and jig bases. A little careful thought beforehand, possibly aided by outside advice such as this article, will enable you to produce a 'third hand' for almost every tricky job on the ship.

With a wooden airplane it is invariably possible to fabricate jigs from scrap timber, proprietary fiberboards and scrap commercial plywood. Nails, screws and glue join together the jig parts. A level surface such as the workroom floor or a large smooth bench top makes the best base for a large jig. Personally, I use my lounge floor with the carpet rolled back!

With a simple fabrication such as a wing rib or laminated bow, the necessary jigs are easy to make and are not in any way involved. However, when a large assembly jig is needed, one golden rule should be adopted. Start on a level surface and work with a level and plumb-bob. If a jig has to be accurate, then the first requirement is that the jig should be level as you will find that by working to a known datum (a level coupled to a centre-line), the job is far quicker and easier.

In this first article, a few hints on wooden fuselages will be given. To enable the reader to understand a jigging pattern for wooden airplanes, it is easiest to detail briefly the construction of a machine wherein particular emphasis will be placed on the necessary jigs. From this basis, a little careful and intelligent thought can convert this to suit whatever project the homebuilder has on hand.

One of the easiest single-place airplanes to build is the Luton Minor and it also serves to employ the basic construction jigs which the amateur will need.

The slab-sided type of fuselage is the easiest to build. Some constructors have built Minors, Turbulents and other types by completing the spruce structure, then attempting to nail on the ply — in space! Unless great pains are taken to level up the flapping spruce frame and keep it free from twist, it is very likely that you will end up with a warped fuselage. Rather like holding a tack still and banging it with the wall to hang a picture up!

First off, Fig. 1 shows the complete Minor fuselage. Make this without a jig of one form or another and you're either a genius or a lunatic! Built in a jig, it's a cinch. Here's how.

Two fuselage sides are needed, basically the same only handed. Should take two hours to lay out the jigs,

Continued on Next Page

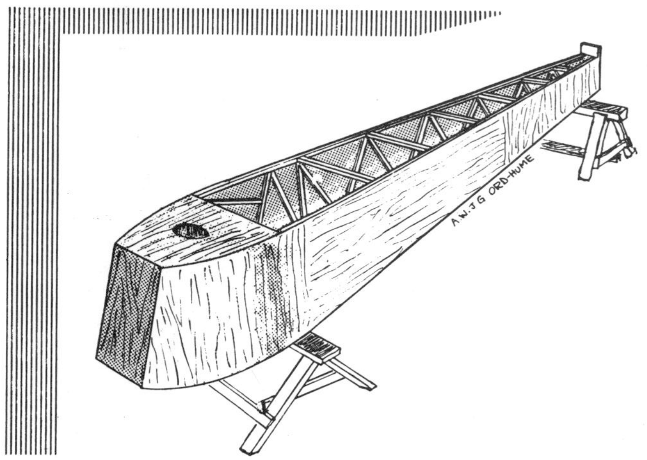

Fig. 1 — Box or slab-sided construction.

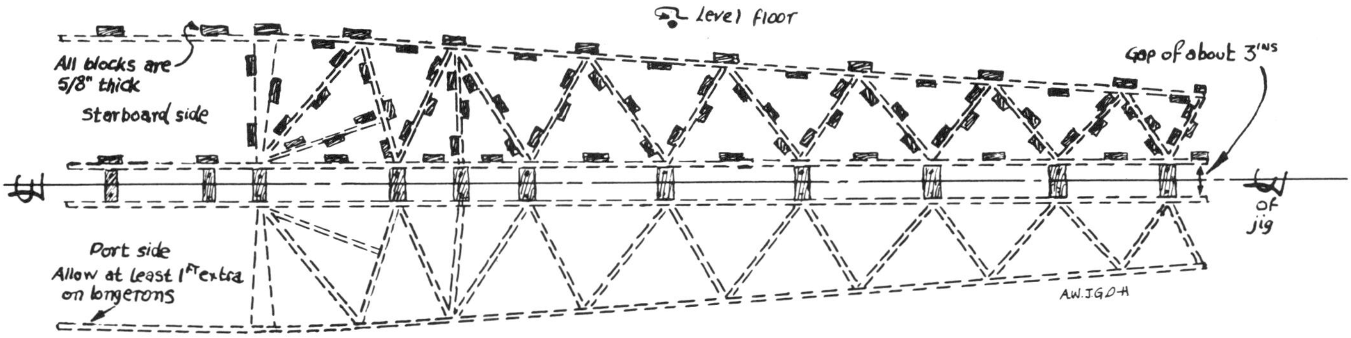

Fig. 2 — The jig for assembling the "Minor" fuselage sides. Note how the blocks are placed at joints to prevent the members bending. In the illustration, half the jig has been completed with the locating blocks. If desired, the members may be located between headless nails in place of wood blocks. However, it is essential to see that the blocks or nails do not protrude above the members otherwise the ply will stand off.

an hour to cut and fit the jig blocks to locate the members, three hours to cut and fit the cross diagonals and longerons in the two sides, an hour to glue up both sides and two hours to fit the ply skin (assuming that it is already scarfed up into one long strip from standard sheets). This shows that the two fuselage sides ready for marrying up can be built almost from scratch within the space of one working day.

With a simple jig for each side, the ply skin is fitted whilst each and every member is positively located and supported by the jig base. If the jig is carefully marked out it is impossible to go wrong. By making the jigs for the two sides side-by-side on the floor, any discrepancies can readily be detected before cutting wood. The illustration Fig. 2, shows the layout of the pair of fuselage side jigs for the Minor.

On this particular airplane, the longerons and cross members are ¾ in. thick varying from ¾ in. to ⅜ in. wide, so all locating blocks should be not more than ½ in. or ⅝ in. thick otherwise they may tend to hold off the ply skin when fixed.

A small length of scrap stock will serve as a gauge for locating the members snugly in the jig. Make certain that the top longerons are straight and not rippled. When all is marked out and checked, bend in the longerons and start cutting the diagonal members. The mitres are easily cut with a small tenon saw or cabinet makers' saw. Clean off any wood burrs before fitting into the jig.

With all the pieces cut, mark each one, remove it and lay it near at hand. Cut a number of small squares of thin plastic sheet (polythene) or waxed paper and place them under the longerons at each joint. This is to prevent glue sticking to the base. Now glue in the diagonals. If they are properly cut in the first place, no additional clamping pressure will be required for setting.

Once the glue is dried, sand the excess off the upper face of each joint and scarf up the ply sheets for the fuselage side skins. Sand the scarfs on the ply smooth before attempting to glue to the side frames. Do not cut the ply to the exact shape of the side frame at this stage; allow at least an inch extra on all sides (plus a good ten inches at the nose) to allow for slight misalignment.

Glue on the ply skin having first marked on the top side the correct positions of all members as a guide for tacking. Apply setting pressure with ordinary staples driven with either an office stapling machine or an industrial stapling gun. Leave overnight to set. If difficulty is exerienced in removing the side from the jig, release all the external locating blocks and gently prise out the side with the blade of a broad chisel. Take care not to crush the wood. Trim off the excess ply, sand off surplus glue on the inside (underneath) of each joint and remove the wire staples with an old ¼ in. chisel.

A useful tip to remember is that plywood attached to a flat frame will after a while slacken and 'oil can'. This is not really detrimental to strength, but it looks very bad and certainly does nothing to improve performance. This can, to a certain extent, be avoided or checked by dampening the upper surface of the ply with a wet sponge immediately prior to glueing it to the frame. This dampening must be done evenly and a few moments allowed for the ply to creep before glueing. When the ply dries out, it will be drum tight. With 1/32 in. thick, ply, it takes only a matter of seconds for the water to soak into the top lamination and expand it for glueing but with ⅛ in. thick ply, ten minutes should elapse. On no account over-wet the ply or allow the underside to become damp otherwise the glue will not stick.

On the Minor, the nose is bent in after boxing up to make certain of the rigidity of the structure and ensure trueness.

Make up four 'L' braces exactly 90 deg. square and screw them to the floor at predetermined stations either side of a centre-line. Between these, clamp the fuselage sides at working height. At all times when clamping to aircraft timber, use scrap wood caul blocks to prevent local crushing of the aircraft timber. The pressure which can be accidentally applied by tightening up a clamp is enormous.

On the Minor it is recommended to construct the fuselage upside down. Level off the two sides at the cockpit and at the stern. Clamp two straight-edged boards along each uppermost edge (Fig. 3) to keep it straight and then cut and fit all the cross-members between the two sides. These are fixed in place with plate gussets and saddle gussets.

Repeat the process for the lower edges (top longerons) and then cut and fit the ply bulkheads and stern post. The fuselage is now locked square but can still warp from side to side.

Make certain that the fuselage sides are in their correct relative positions and that a line drawn between two corresponding points on each side cuts the centre-line at right angles.

Fit the bottom plywood skin from the stern post to the edge of the forward ply bulkheads (Fig. 4). When nailing or stapling the ply to cross members between the sides, have an assistant dolly up the member from underneath with a heavy weight or the flat of a heavy hammer otherwise the wood may split and the terminal glued joints will crack.

Turn the fuselage the right way up and cut and fit the diagonals in the top side between the cross members having first clamped the sides between the straight boards

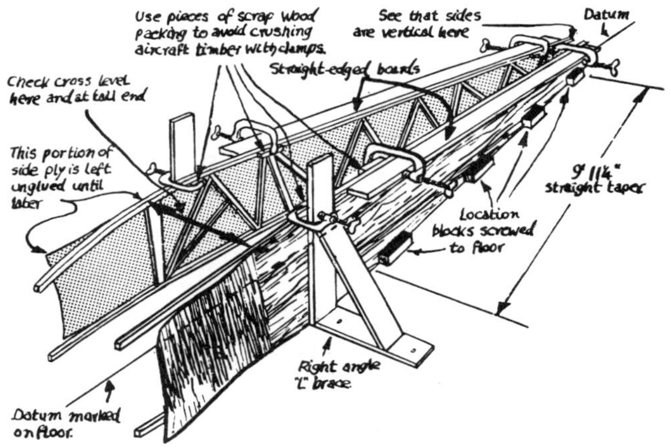

Fig. 3 — View of sides (inverted) clamped between straight-edged boards.

again as before. If this is not done, the longerons can still be warped from side to side, the effect being to produce a lozenge cross-section. Once the top cross bracing is in, the fuselage is locked perfectly rigid aft of the cockpit to the stern post.

Bending the longerons at the nose is a difficult job unless tackled with a jig. Unlike steel tubing, wood is springy and resilient and, even when steamed, will spring to a certain degree. Unless the bend is acute (and it is not likely to be), it can be bent in a three-dimensional jig. On the Minor it will be seen that the ply has not yet been fixed to the nose longerons. This is because if it were to be glued up it would be very

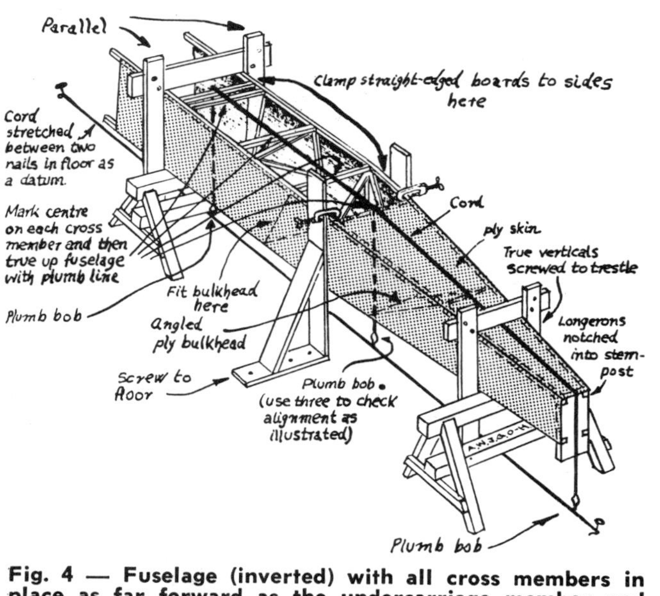

Fig. 4 — Fuselage (inverted) with all cross members in place as far forward as the undercarriage member and main bulkheads with stern post fitted.

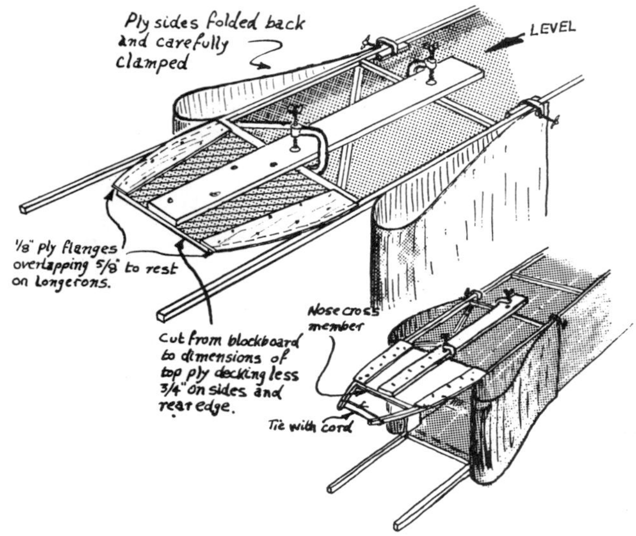

Fig. 5 — Jig for bending in top longerons. This is left in place until lower longerons have been bent, the engine bulkhead fitted and the ply side skins glued on.

difficult - if not impossible - to bend in the fuselage sides due to the tension of the ply.

Fig. 5 shows a simple two-dimensional jig for bending in the top longerons which are formed before tackling the lower ones.

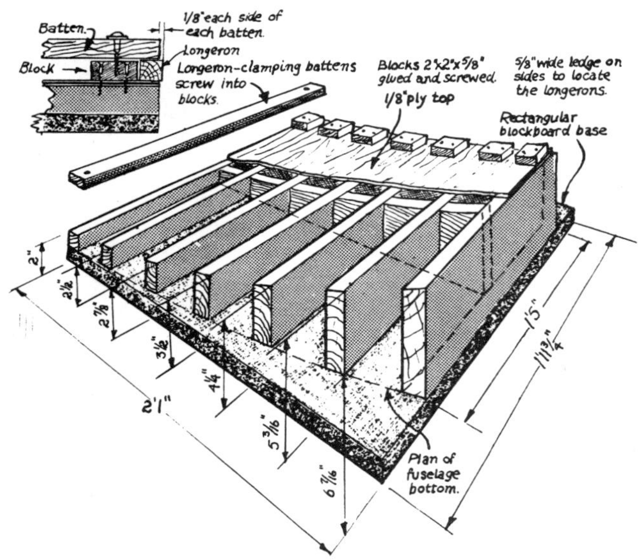

Fig. 6 — Jig for bending bottom longerons.

A three-dimensional jig (known sometimes as an 'egg-box') is easy to make and is, of course, standard practice in the manufacture of sheet metal compound shapes such as cowlings and so forth. Again, always work from a datum or level. Fig. 6 shows the three-dimensional jig for the lower longerons and Fig. 7 illustrates the jig in use.

By arranging the lengths of the jig members and cross battens so that clearance is allowed on the sides between the edges of the jig members and the longerons, the side ply skins may be attached and, in the case of the Minor, the engine bulkhead assembled before removing the jigs, thus ensuring that everything is held rigid and in place until boxing up is completed. A word of warning, however, on three-dimensional jigs. Unless the members being bent are steamed (this is not always essential) there will be a tendency for them to twist outwards so

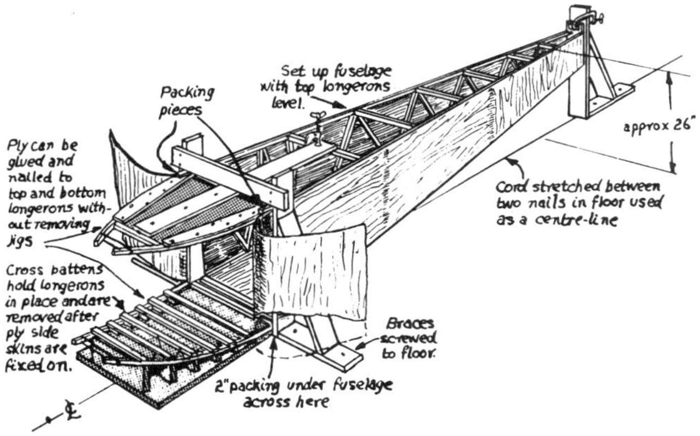

Fig. 7 — Fuselage set up for bending in bottom longerons.

that their horizontal faces incline outwards. This twist must be resisted in the jig - in Fig. 7 the twist is counteracted by the cross battens of the jig.

The Minor has a stiffening former of laminated ash glued diagonally inside the cockpit and this is fabricated **in situ** having first clamped the cockpit sides between the boards and ensured that the sides over the cockpit length are parallel.

This completes the basic jigging of the slab-sided type of wooden fuselage.

A monocoque fuselage is one which has no longerons, the outer skin absorbing all loads, normally in torsion. The monocoque fuselage is usually circular, oval or egg-shaped in section and the skin is stiffened with transverse formers or hoops and longitudinal stringers.

This type of fuselage is more difficult to construct and requires a different approach regarding jigs. The easiest method is possibly the beam jig to which the frames are fixed and lines up. The stringers are then fixed in their notches in the formers and the ply covering applied. It is a good scheme (where possible) to leave the bottom of the fuselage open and clear in the jig until the remainder of the fuselage and stringers and the major part of the skin are in place so as to allow ample access to the inside. If practicable, it is an advantage if it is possible for an assistant to stand inside to dolly up stringers for any tacking which has to be done.

If possible the jig should be made to rotate longitudinally to allow access to all parts of the fuselage during construction work. Typical jigging arrangements for this type of construction are shown in Figs. 8, 9 and 10. When the fuselage is completed, the jig may be released and slid out from inside the fuselage. Alternatively, the beam may be built up of bolted alloy section and dismantled inside the completed fuselage.

Another method which is suitable for even-tapered fuselages, nacelles or booms of parallel or frustrum section is the female jig which can be made of fiberglas or plaster. In this, the skin is put in first and then the internal structure built in. The two halves are then joined together. This type of jig is only suitable for certain applications, although it is ideal for large cockpit canopies, detachable curved panels and similar assemblies.

One final word for the moment on jigs. Don't improvise trestles from old crates and boxes. If you have any ideas about accurate lining up and levelling, use good solid pieces of timber for trestles and primary jig members. Remember that money spent on jig materials is not money wasted — it can save you from wasting costly airplane material and can save you plenty of time

In a further article, jigging for wings will be detailed and the importance of space-jigging explained to those who may be unfamiliar with this procedure.

Fig. 8 — Monocoque Construction.

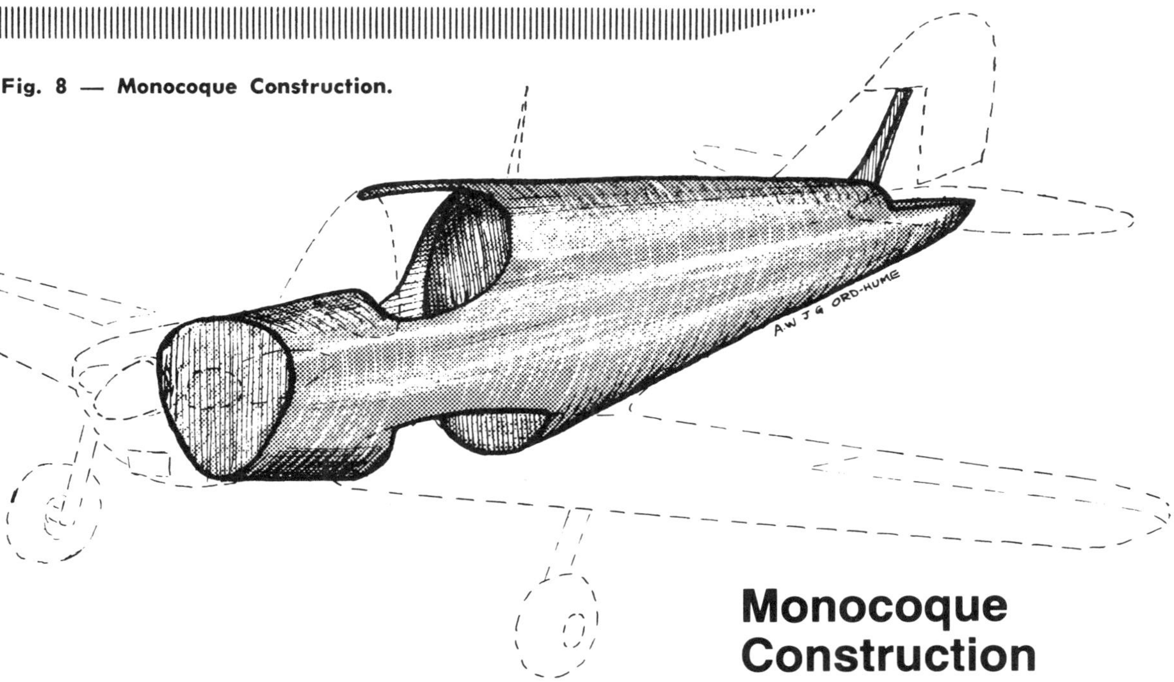

Monocoque Construction

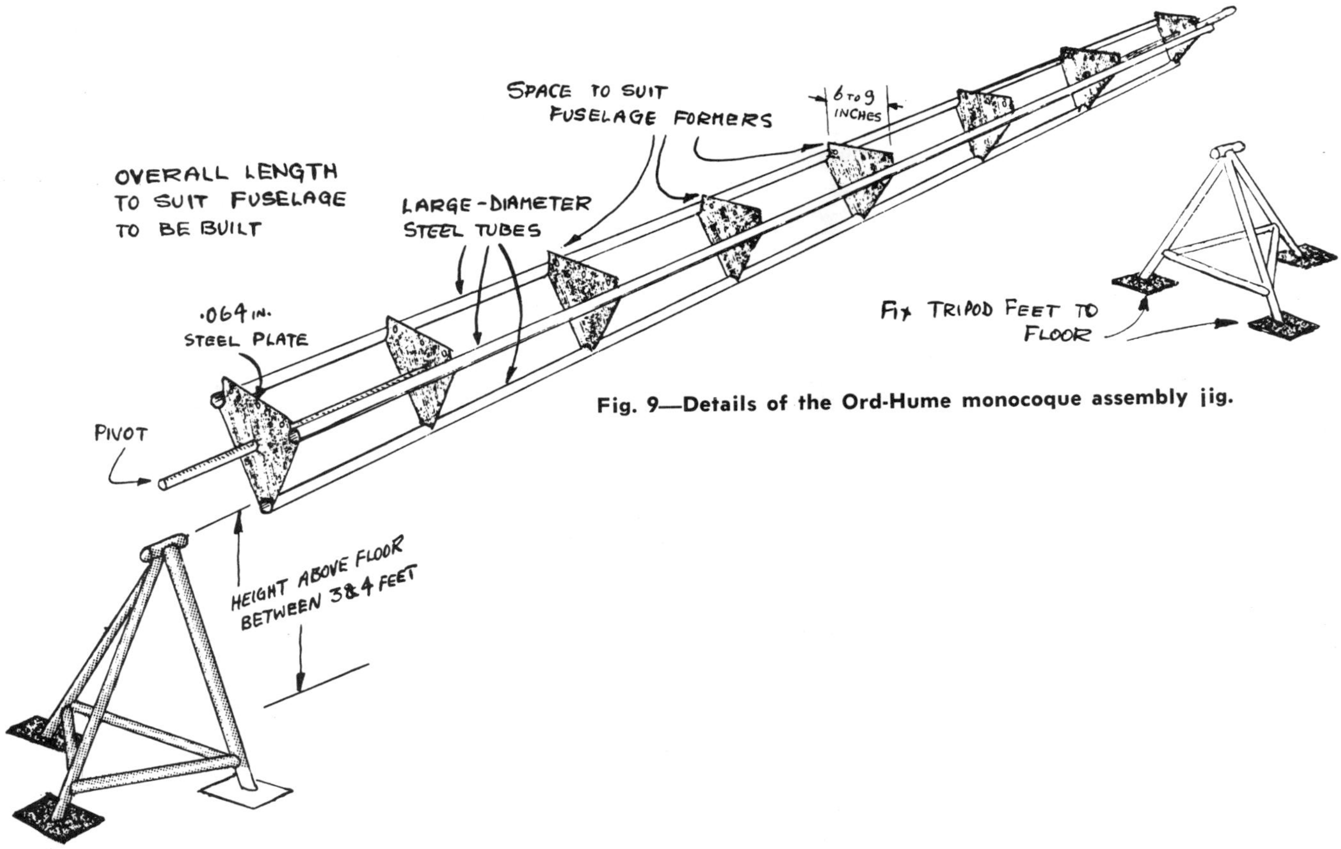

Fig. 9—Details of the Ord-Hume monocoque assembly jig.

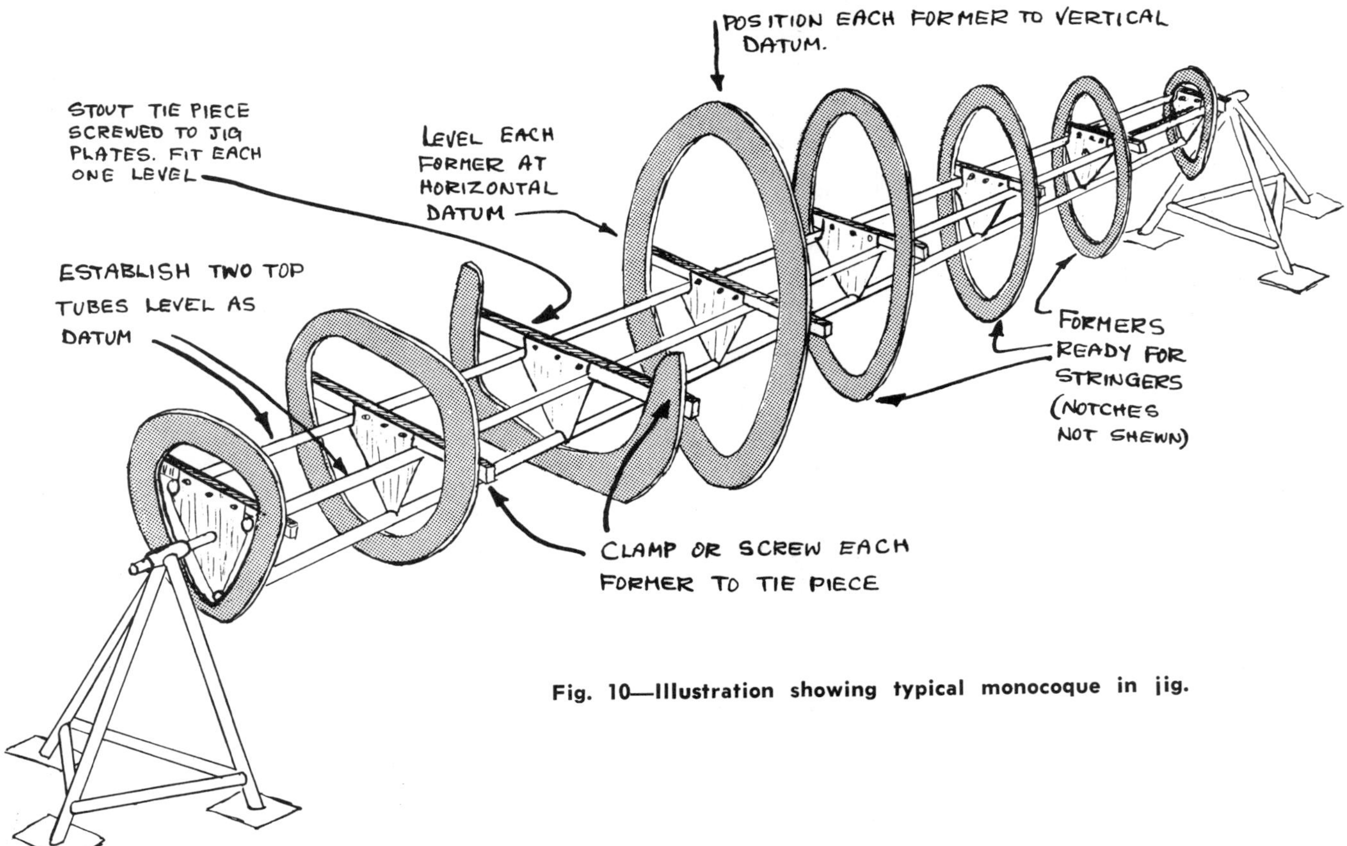

Fig. 10—Illustration showing typical monocoque in jig.

Part 2

Jigs for Wooden Aircraft

Wings

By Arthur W. J. G. Ord-Hume, EAA 8579

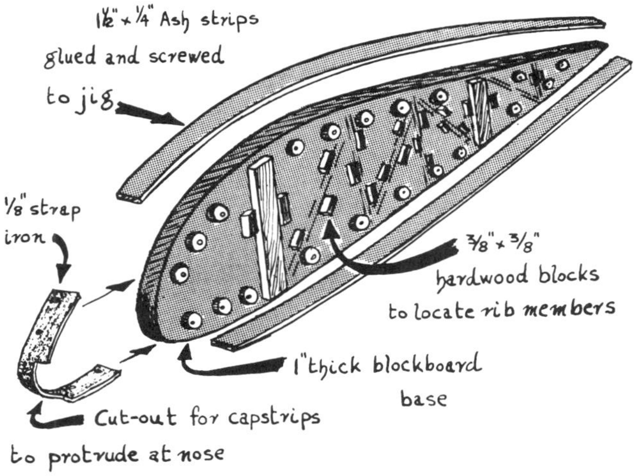

Fig. 2 The Ord-Hume improved rib jig for rapid rib construction. Paint with two coats of hot linseed oil to prevent glue adhesion. Make certain that the oil is thoroughly absorbed by the jig before use.

As in the case of the fuselage, jigs play an important part not only in assembling the component parts of the wing such as the ribs, compression struts and spars, but also in the correct alignment of the complete panel prior to skinning the leading edge with plywood, or, in the case of a stressed-skin wing, the covering of the entire wing.

The same rules apply to wing jigs as those detailed earlier in Part 1 of this article. Work from a level, establish a datum and a centre-line which may be represented by a line on the floor or a line in space represented by a taut string. Check each and every layout — if possible have a friend check with you. Never rely on purely eyeball estimation — a surface or component may **look** level but, until someone devises a calibrated graticule with vernier adjustment to fit your headlamps, stick to to the old-fashioned adjustable bubble level and straight-edge.

About the first jig (often the only jig!) one thinks of with a wing is the rib jig. This is a peculiarly satisfying piece of carpentry — indeed almost as satisfying as the ribs which are prised out of it! The form of the rib jig has not changed over the years but I have devised an alternative jig which I find a little easier to make and use. To begin with, Fig. 1 shows the conventional jig. Fig. 2 illustrates my jig and here's how to make it. Draw the full-size profile on to a plank of laminated block-board about one inch thick. Mark the profile very clearly — the plank may first be painted white after sanding and the outline marked in India ink. Now take the board to the bandsaw and rip out the shape, cleaning down to the profile with block-plane and rasp. Cut next a number of wood buttons about one and a half inches in diameter and slightly thinner than the width of the rib stock. With practice, these buttons can be cut out with a tenon saw as octagons and then spun between the fingers against a sander to circular shape. Drill each one 3/16" diameter ⅛" off centre. Assuming the capstrip stock to be 3/16" thick, the button should pivot approximately 13/16" from the outside profile of the rib. Scribe off this distance parallel to the edge of the board with a pair of odd-legs. Making sure not to interfere with the bracing members of the rib structure, drill through the board to take 3/16" bolts at each button. Bolt on the buttons with a washer under the bolt head and another between the button and the jig. Tighten sufficiently to hold the button yet permit it to rotate reasonably freely. Plane up a couple of pieces of scrap wood to the cross-section of the spars and glue and screw these into place. Now glue and tack on pieces of rib stock to locate the cross-members and diagonals.

Two strips of ash 1½" by ⅛" are now glued and screwed to the upper and lower edges of the jig as shown. Because the spring of the capstrip over the sharply-curved nose of the rib might split the ash profile re-

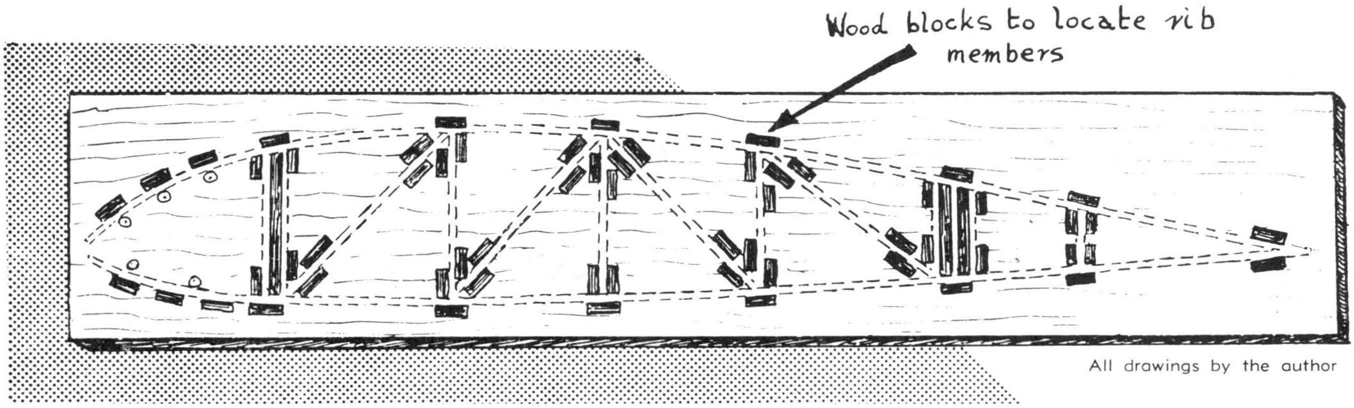

All drawings by the author

Fig. 1. Conventional rib jig laid out on base board.

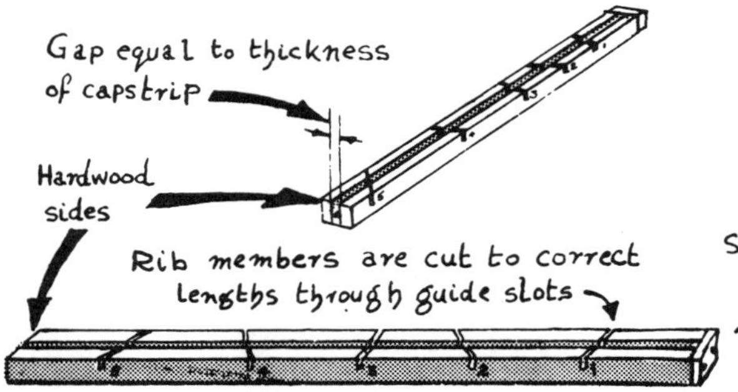

Fig. 3. Jig for cutting rib members to length.

JIGS . . . *from preceding page*

tainers, the first few inches of the nose is clad with a strap iron profile. Clearance is made in this for the capstrips to protrude before being trimmed back to size.

A word here on rib building. When the rib is glued up and the gussets put on the top side in the jig, the rib can immediately be lifted out, turned over and the gussets fitted to the other side. Each joint can therefore be checked at once for alignment before the glue dries. To make this possible, the jig should be used on a clear surface with sufficient flat space by the jig to work on the rib after its removal from the jig. Alternatively, if space is at a premium (it certainly is in my attic!) find a large board big enough for the rib to be supported on and drill and fit this with two ½" diameter dowel pegs to recess into the jig itself. The rib may then be taken from the jig, the flat board keyed on top of the jig, and the rib worked on on the clear space provided.

A job which takes time and much fiddling can be the cutting of the rib bracing diagonals and cross-members. The guy who cuts each one individually for every rib is a mug! Spring two capstrips into the rib jig, press them out against the profile retainers by means of the eccentric buttons and carefully cut and fit a full set of diagonals and cross-members. Number each rib piece in the rib on the jig. Take two pieces of good straight commercial timber about 1" by ½" and glue these either side

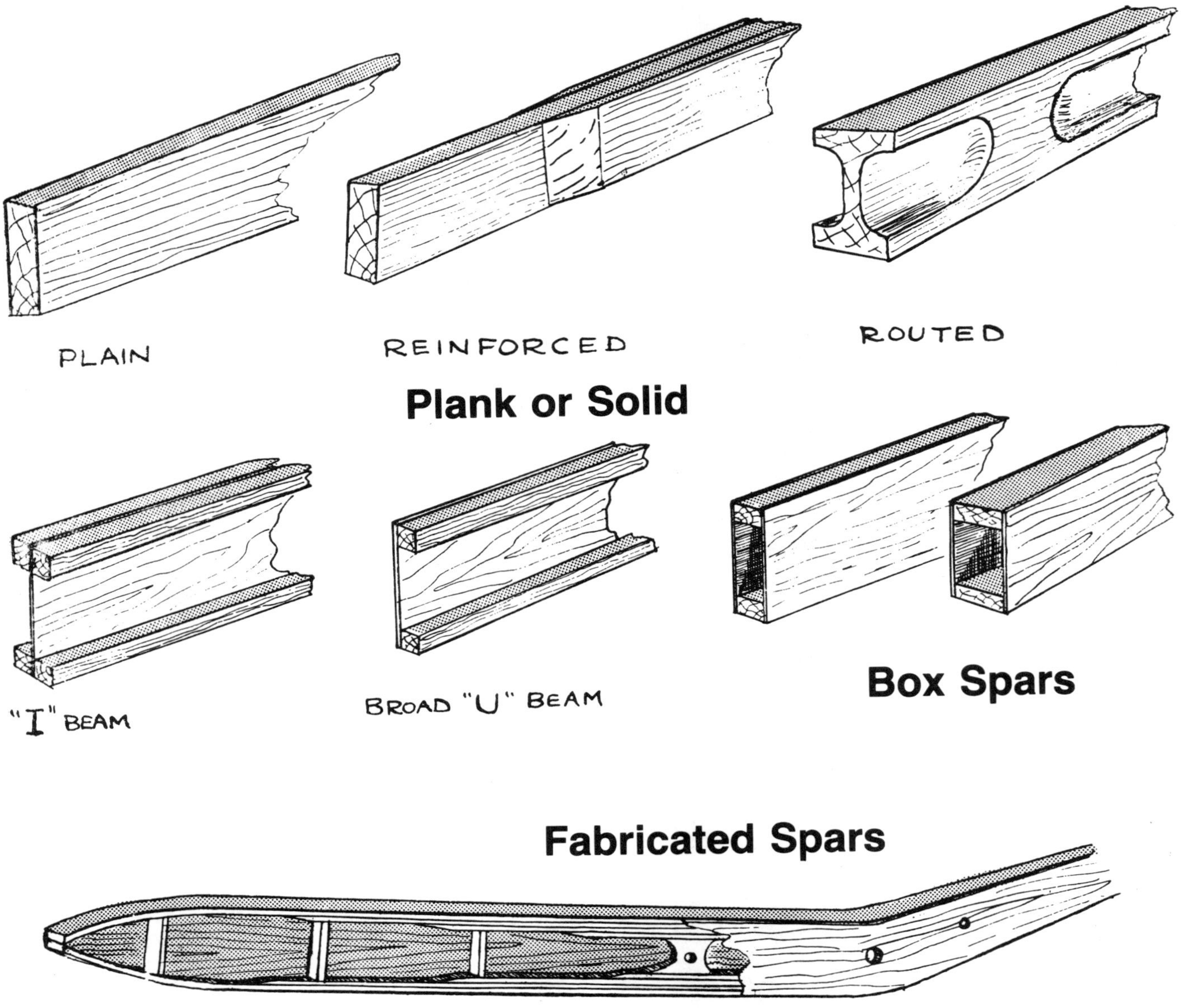

Fig. 4. Types of wing beam or spar.

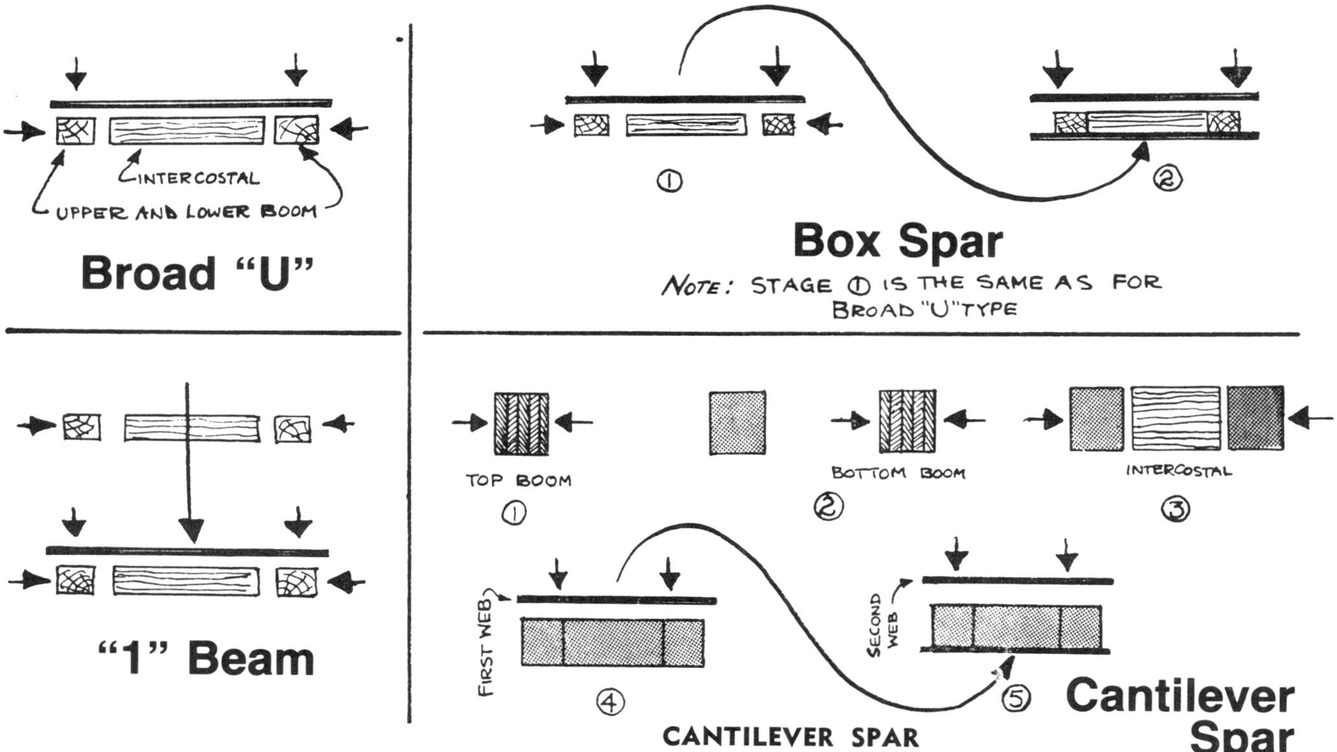

Fig. 5. Step by step stages in the construction of types of built-up spar. The arrows indicate the direction of clamping pressure at each stage.

of a spare length of rib stock (Fig. 3). The rib members are put into this end to end and the assembly converted into a mitre block with a small tenon or cabinet-makers saw. In the illustration, only five members are drawn. I have actually colour-coded my own jig, members 1 to 5 marked in red, 6 through 9 being blue and 10 to 14 green. All these members are cut in one mitre-block only three feet long and the only point to watch is that only one colour is cut at a time, the pieces being removed, the capstrip slid back up to the stop again and then all the next colour cut. This device technically comes under the heading of a tool rather than a jig but the amateur need not worry what to call it — if it's of use, then use it whether it's a wimwom or a whatsit.

SPARS

There are four types of spar commonly used in light aircraft. First is the plank spar which is usually of solid spruce, sometimes routed out to save weight. Next comes the simple box section spar and its variants wherein a single straight top boom and a similar lower boom form the upper and lower sides of a ply-sided box. The third type is similar to the second, but the spar has a single ply web with spruce booms fixed to the top and bottom to make a broad 'U' section, or a straight 'I' section.

The fourth type of spar is the complex laminated cantilever spar for a one-piece wing (in the case of a glider a two-piece cantilever wing) which comprises a top and bottom laminated built-up cantilever boom, boxed together with plywood.

These types are illustrated for reference in Fig. 4 and each presents jigging problems of its own.

The solid spar is the easiest of them all — it is planed up straight and rectangular and that's it. Some airplanes have plank spars which are strengthened locally at high stress points by plank doublers, an example being the Aeronca C-3. Since it is well known that plank spars are unnecessarily heavy, the central part of the plank acting purely as a high weight penalty shear web, many airplane manufacturers make use of routed plank spars wherein the 'dead' wood is cut away to a thin web. An example here is the de Havilland Tiger Moth (indeed, **every** type of Moth employed routed plank spars!) Routing can be done with a routing cutter in a pillar drill, but that is outside the scope of the present article.

The simple box spar and the 'I' or broad 'U' beam have certain points in common so far as basic assembly and jigging is concerned, so we will look at these together.

Fig. 5 shows the sections of the spar types and illustrates the directions in which setting pressure must be applied. Pressure direction should be considered early on in the design of the jig as often jig members have to be arranged to provide clearance for, or location for clamps, sash-cramps or pressure fixtures.

It will be seen that the section of the spar as drawn shows the ply webs protruding beyond the line of the spruce booms at each side. This is to allow for trimming afterwards. Never attempt to cut a piece of ply dead to size for fixing to a frame, fuselage side or spar for the simple reason that it is very difficult to ensure correct re-alignment in assembly and it is far easier and safer to allow a 'flash' to remain which can be planed off afterwards.

This flash must be allowed for in the jig design and some idea of this type of jig is shown in Fig. 6. Note that the intercostals in the spar itself are glued in before the second boom is positioned and fixed. This is because if they are glued in afterwards, inserting each piece into a gap which is dead to size will clean all the

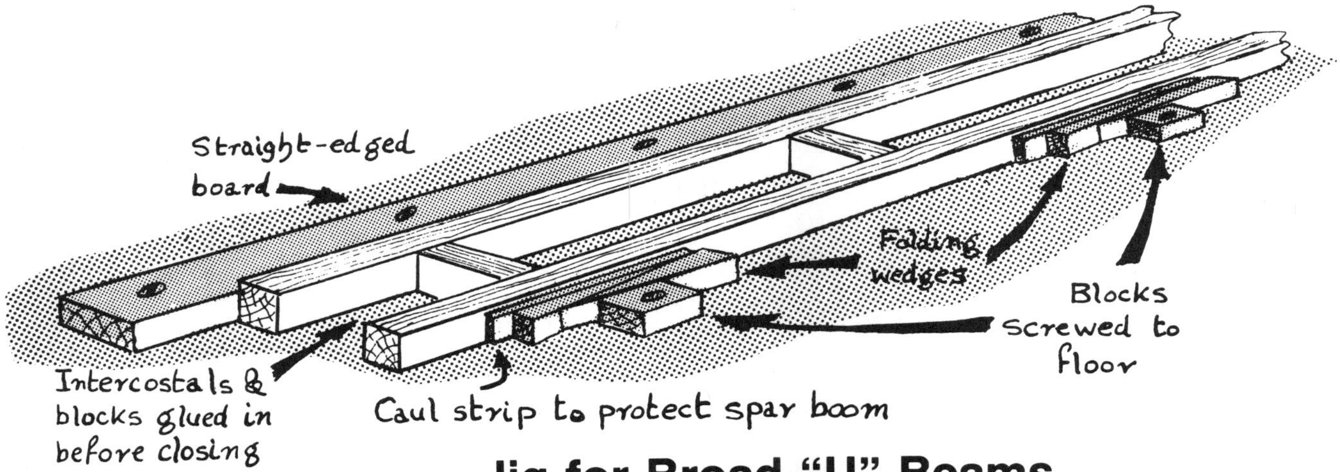

Jig for Broad "U" Beams and First Stage for "1" Beams

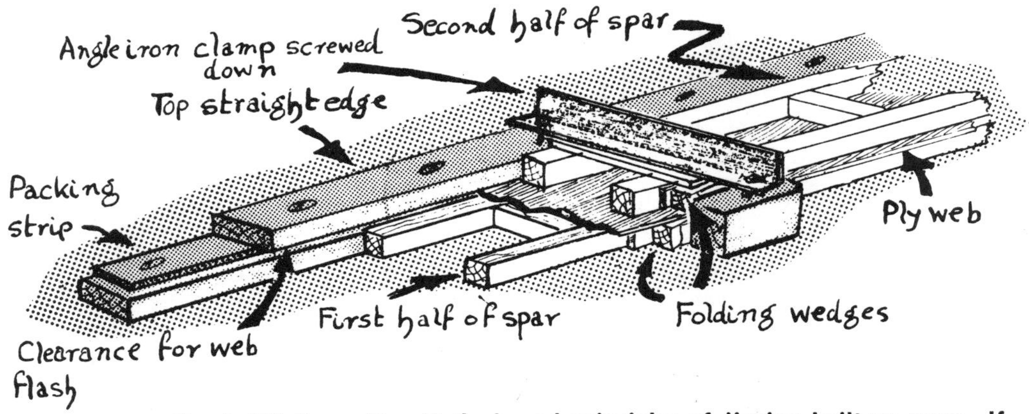

Fig. 6. "I" Beam Jig. Methods and principles of jigging built-up spars. If a spar table is used, "C" clamps can be used in place of angle iron clamps.

JIGS . . . *from preceding page*

glue off. The second boom is closed in with folding wedges.

For a box-spar, fit the booms and intercostals first and then the uppermost ply skin. When thoroughly set, remove the spar very carefully and fit the ply skin to the other side. Similarly, when building a broad 'U' beam, arrange the jig so that the spar ply web will come uppermost. In the case of 'I' section beams, a two-stage jig is used as shown in the illustration.

Spars may be built on a wooden floor in which case clamping pressure is obtained by screwed clamping pads made either of hardwood or of angle iron. Ideally, of course, the spar should be built on a spar-building table, detailed in Fig. 7, clamping pressure being achieved with C clamps or toggle clamps.

The complex cantilever spar requires a very different technique and, since it is imperative to work at a convenient height rather than on the floor, make up a spar-building table as shown in Fig. 7. The tops of the frame trestles are levelled off and boarded with thick planks of wood which are then planed smooth. Arrange the profile of the jig table to that of the spar and then make the table only a few inches wider than the spar so as to allow C clamps to be used.

Begin by laminating up the actual spar booms, first the top, then the lower. When all the various laminations are prepared for gluing, have a 'dummy run' in the jig and arrange beforehand just where the clamps and blocks are to be fixed. Also open up the clamps to the right gap and have immediately to hand everything necessary for assembly because, once the process of gluing starts, all clamping must be completed within the life of the glue. Recruit a couple of friends and run through the procedure with them. Alert the wife (or similarly convenient standby person) and make certain she understands how to answer doorbells, telephones and so forth, for once you start gluing, nothing must be allowed to stop you until the boom is clamped up.

A word of warning. Never buy your spar lamination stock planed to dead width. Always allow at least one eighth of an inch extra on width to allow for shuffling and cleaning down to dead width with the smoothing plane **after** laminating.

As throughout all clamping operations, clamp on scrap pieces of timber to protect the aircraft wood from being bruised. There is an old cabinet-makers' stunt of wetting a bruise on wood so that the collapsed wood cells expand and hide the bruise from sight. Very good tip to remember — when making a bookcase! Forget it in aircraft work — the collapsed cells are still weakened and useless whether you spit on them or 'accidentally' blob on a lump of glue. The best way is to prevent bruising right from the start.

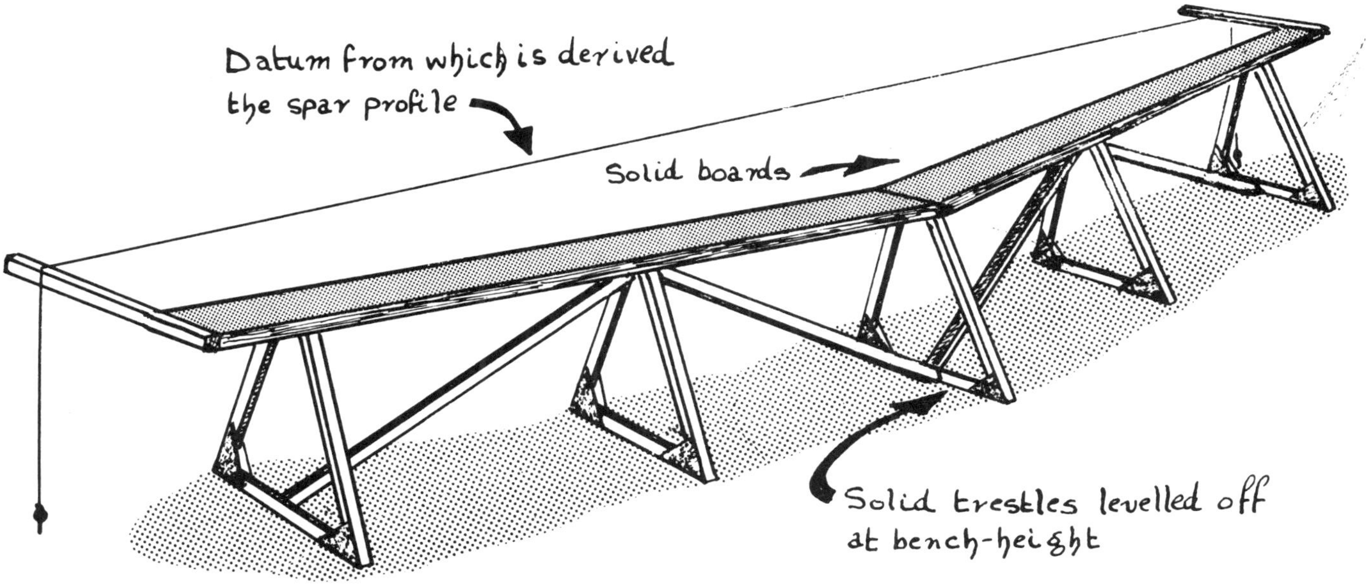

Fig. 7. The layout of a spar-building table. The number of trestle supports depends on the length of the spar as the working surfaces must be level and solid.

PANEL ASSEMBLY

The complete wing panel can be put together on two saw-horses, all ribs, compression struts, fittings and internal brace wires (if used) being positioned, the last-mentioned being left loose for the meanwhile. Whether the leading edge is to be skinned with ply, or the whole panel ply covered, accurate alignment in a jig is essential. This is because the moment the ply is fixed in place, the wing becomes a completely rigid panel and any twist, warp or lozenging accidentally occurring will be set in, attempts to graunch it back into shape on the ship by heaving up on the lift-struts only serving to prestress the whole structure.

The best method to support the panel is to hold it vertically with the leading edge uppermost. Jigging supports are shown in Fig. 8 and the process of alignment clearly outlined in Fig. 9.

For a fully ply-covered panel, cover the **lower** surface of the wing first, jigging by clamping on the spars and the root ends. The wing is then removed, turned end over end (so that the leading edge is still uppermost) and re-jigged picking up on the strut attachment brackets (protruding through the lower ply) and the root ends. In this manner, positive alignment can be ensured.

An important point to watch is that all ribs are profiled before ply covering after jigging the wing using a straightedge and checking for high and low spots. The slightest irregularity between capstrip heights can spoil the looks of a skinned panel as well as making it extremely difficult to get the ply down in contact with the rib. On high-speed airplanes, the resultant bubbled skin can knock off mph at the top and, sometimes overlooked, can raise the stall. Broadly speaking, the surface condition of the fuselage doesn't matter much. It is the wing wherein can lie the success or failure of a ship, so it pays to take pains with skinning.

When skinning a wing, back up the trailing edge of the ribs with a good stout straight edged board clamped along all the ribs. This will prevent waviness of the edge as well as making a far more solid structure to hammer or staple against.

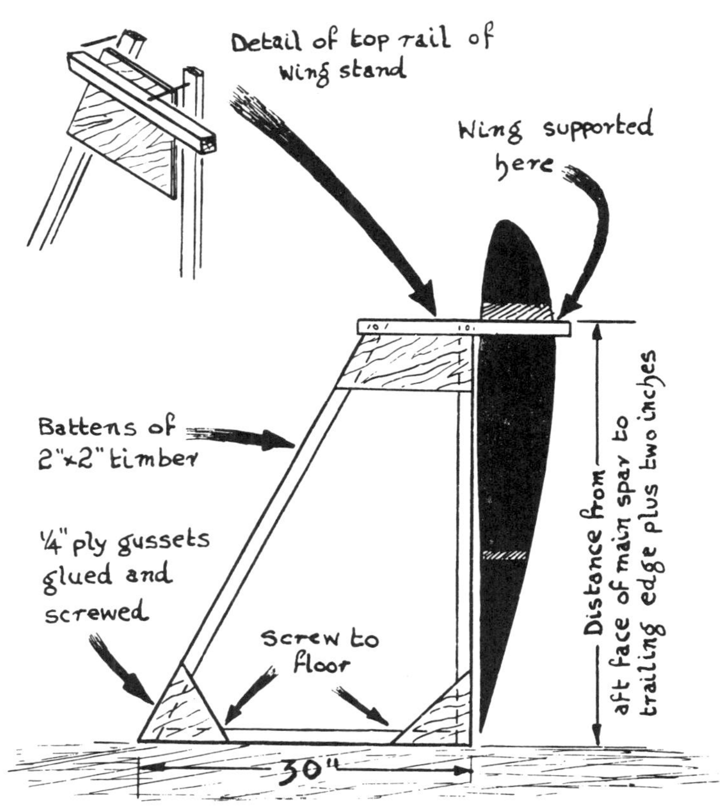

Fig. 8. Detail of the wing supporting jigs for skinning. These can be made of wood or slotted steel angles.

SPACE JIGS

A recent innovation in house construction has been the odd approach of putting the plumbing in first. As long ago as 1948, houses were built in Canada and elsewhere starting as foundations, rising water main, water tank, cisterns and basins. Must have looked odd to see a site of concrete bases and swaying pipes and sinks high in the air. Enough to make a fellow sign the pledge on

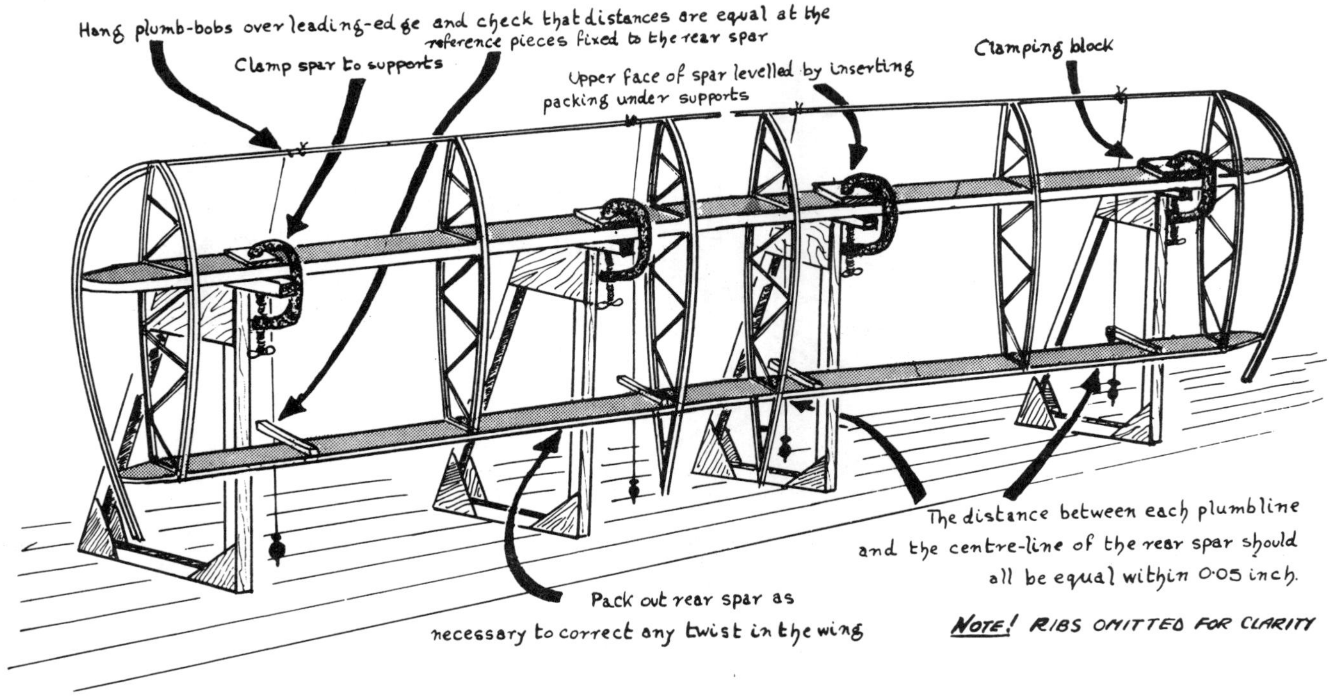

Fig. 9. Panel in place on jig for skinning. With the frame rigidly clamped at the main spar, twist can be measured by the plumb lines at the rear spar. Packing between the rear spar and the jig support will remove any twist. Method of jigging is suitable for both one-piece and two-piece mainplanes.

the spot. Nevertheless, the architect boys had a notion which paid off with several advantages.

Same with airplanes — although no homebuilt has yet appeared with a low flush suite. The idea is simple. The best application is the parasol monoplane wherein the builder sets up his fuselage on trestles, supports the wing over the top at the correct incidence and then cuts his pylon tube to fit between the two.

Often, a fitting has to be in a special, predetermined place in a structure, tied, perhaps, to a datum and an elevation. Rather than fiddle it into place, put it in first and build the supporting structure up to it, thus ensuring that it is in the right spot.

Reverting to wing pylon fittings. Set up the fuselage level, hook up a string centre-line parallel to the top longeron and, from a plumb-bob and square, calculate the position of the front spar. Now make up a timber or angle frame fixed to the fuselage or, if the fuselage is positively fixed to the floor, fit the frame also to the

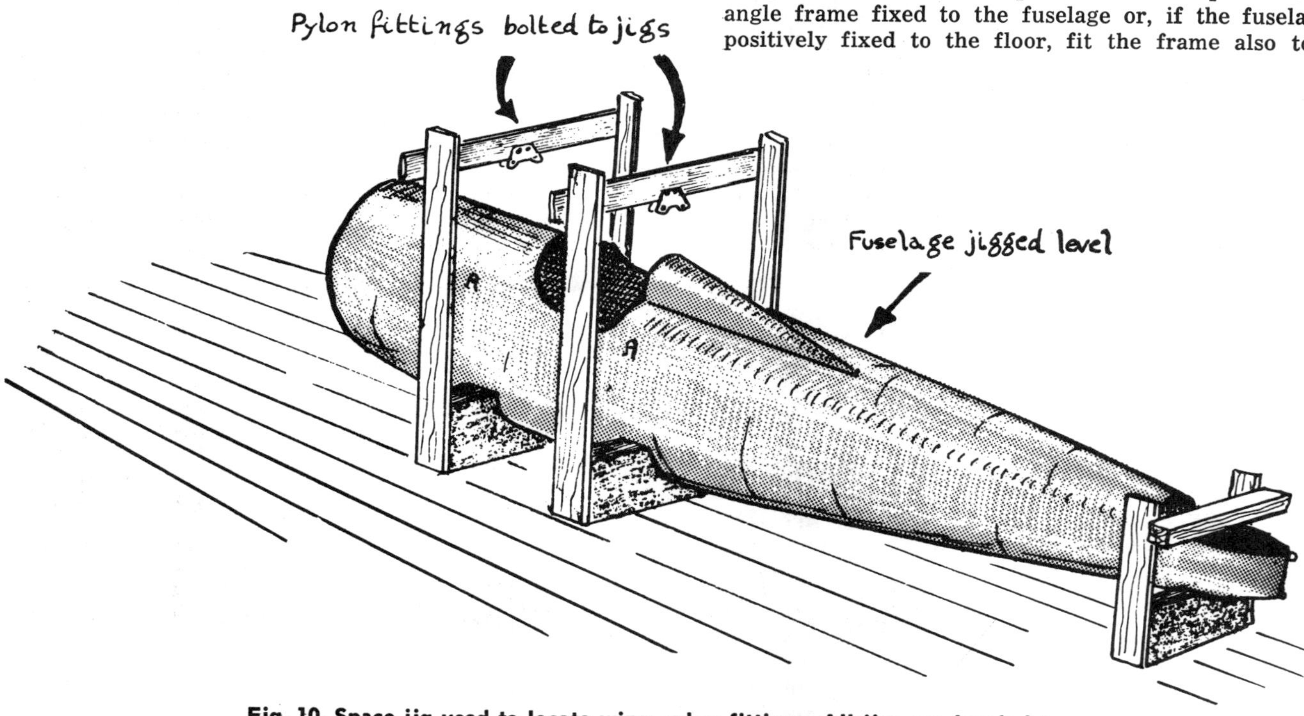

Fig. 10. Space jig used to locate wing pylon fittings. All jigs are leveled to the common level at the fuselage and are spaced from the fuselage datum.

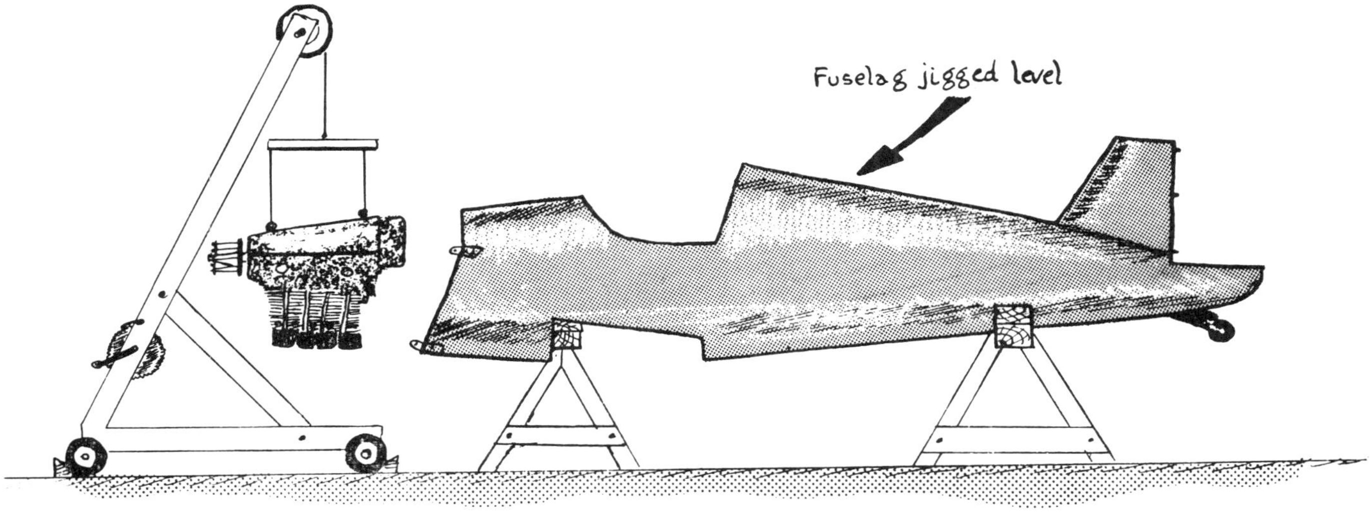

Fig. 11. The space jig principle adapted to the making of a motor mount. The fuselage is jigged up level and the motor (or representative jig) positioned correctly. In the illustration the motor would be rigidly wedged against movement and leveled from the engine datum to the fuselage datum.

floor. To this frame, bolt or clamp the front spar pylon fitting. Now find the distance between the spars and the dihedral angle. Calculate via trigonometry the drop of the rear spar fitting (tan. [dihedral angle] $\times$ distance between spar fittings) and make a frame to hold this one. Fix the two frames together to maintain the between-spar distance (Fig. 10).

Tubes can now be cut and welded or bolted to suit between the two. The supporting framework is the space jig insomuch as that it is firmly anchoring a point in space. The same method holds good for making up engine bearers when the motor crankcase (or jig fixture resembling it) is set up in place in the correct relationship to the fuselage and the mount schemed between the two. (Fig. 11).

The space jig method is ideally suited to experimental and prototype work, although the principle can often be employed with a regular design. Some wise guy is now a-painting his licence number on a piece of rag and reckoning on building his airplane behind it!

The whole essence of aircraft construction is a careful, intelligent approach. The same is valid for jigging. A lot of time can be wasted in the workshop scratching and wondering what to do next, or how to do something. Don't wait until you are standing in the shavings to ponder—fathom out each problem whilst going through the normal daily routine. My most valuable construction hours are those accrued in odd moments in the office, waiting for connections and whilst travelling, for it is then that ways round knotty problems may be found and the next construction stages planned.

Of all the materials available today, wood is about the most satisfying to work with and with modern varnishes and synthetic resin glues, the wooden airplane is as robust and as durable as almost any metal ship. It is not so many years ago that engine bearers of wood were common and I know of one very successful home-built which first flew twenty-five years ago and still flies today with its original wooden engine mount.

The durability of a wooden ply-covered structure can greatly be enhanced by covering the wood with madapolam cotton fabric, doped on after two coats of primer dope. This also makes a sound base for building up a first class gloss finish.

In France and England, aircraft tube is very expensive and delivery dates of twelve months are common. Although this certainly affects the selection of wood as a prime construction material, there is still the very personal gratification of planing, sanding and glueing this, the most versatile of all materials.

An extremely simple all wood aircraft to build is the Evans VP. Many of these aircraft have been built as the VW powered VP permits minimum investment for materials and is very economical to operate. Note flat sheet plywood fuselage. Couldn't be simpler!

Building Wooden Cantilever Spars

By Arthur W. J. G. Ord-Hume EAA 8579

SINCE PUBLICATION of the earlier series of articles on JIGS FOR WOODEN AIRCRAFT, a number of members have written asking for detailed steps in the construction and assembly of cantilever spars or beams, starting from raw wood through to the assembly of the actual wing. Reference to Parts 2 and 3 for the previous series of articles should be made (SPORT AVIATION, January and February).

A cantilever spar comprises two built-up laminated booms, normally an upper boom corresponding to the top surface of the wing, and a lower boom corresponding to the lower surface of the wing. Both these booms may comprise several thicknesses of spruce planks or strips bonded together by glue.

At the center of the cantilever one-piece wing the stresses are highest and therefore the individual spar booms are thickest, a relatively large number of separate planks or strips being bonded together to provide large cross-sectional area. As the spar proceeds outwards towards the tips, the stresses diminish and the number of thicknesses of wood in the booms is reduced. These reductions must not be by means of square-cut steps which would create a high stress concentration at the remaining wood layers, but must be smooth tapers akin to scarfing (Fig. 2).

The upper boom takes a considerable compression load in normal flight and maneuvers; the lower boom is in tension. Even for inverted flight sometimes a large proportion of the load reversal can be accounted for in the upper boom in the spar and its shear webs to the lower boom without excessive strengthening of the lower boom for the inverted flight case. From this we see why it is that the upper spar boom usually has much more wood in it, represented by a larger cross-sectional area or more wood laminations, than does the lower boom.

Earlier, reference was made to the smooth diminishing of the individual laminations towards the tip. Because of the balance of the stresses in the wood of a properly designed spar of a given strength without excessive weight, any joints between portions of the individual laminations in the beam must be carefully made scarfs.

Furthermore, if it is necessary to scarf-joint the individual laminations (which should be avoided if at all possible) the joints must be staggered so that they do not come adjacent or superimposed upon similar scarfed laminations in the boom (Fig. 3).

The two booms in the spar are separated by one or more ply shear webs as shown in Fig. 4 on page 16 of January SPORT AVIATION. Because of the shear tendency between the spar booms is maximum along a line more or less along the physical center-line of the spar as a whole, the grain of the outer veneers of the plywood is normally spanwise and not vertically placed, compression between the booms being taken care of by vertical "soldiers" thereby reducing the individual ply panel sizes. There is, however, no definite rule on this point and the particular design being built will specify the grain direction of the plywood and this must be followed.

To prevent the booms crushing the ply webs and to provide strengthened points for the attachment of struts, landing gear or fuselage attachment points, various intercostal members and blocks are glued in between the upper and lower booms. These serve also to reduce the ply panel area as described above. These vertical members are normally placed to coincide with the point of attachment of the ribs so they act also as a surface stiffener to which may be glued and tacked the vertical members of the rib which are adjacent to the spar.

Having dealt with the basic items which comprise the cantilever spar, let's look at the problems involved in making it. First of all comes the job of marking out some form of accurate layout to which the spar may be built. From this the jig has to be made. Next comes the assembly and glueing up of the various laminations, bearing in mind the fact that

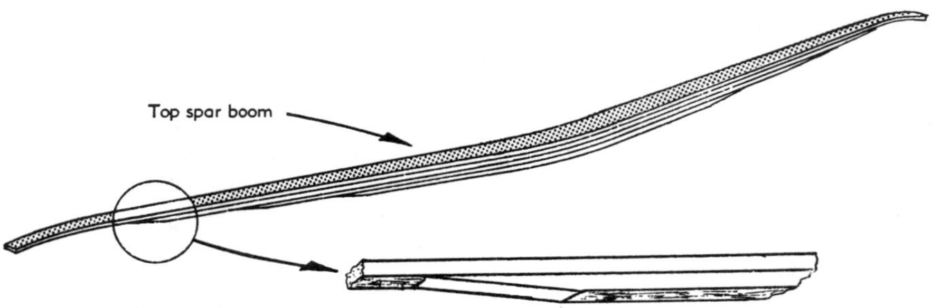

FIG. 2 The ends of each lamination must be feathered to avoid stress concentration at what would otherwise be a sharp change in cross-sectional area.

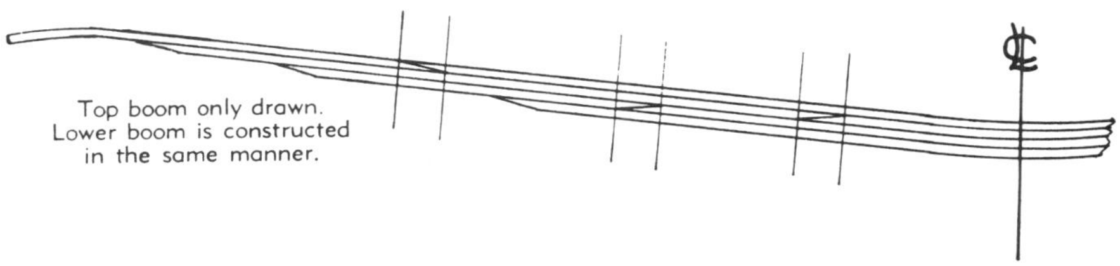

FIG. 3 If it is necessary to scarf individual laminations, stagger the joints so they are not superimposed on each other. Do not scarf laminations near center section or landing gear attachment.

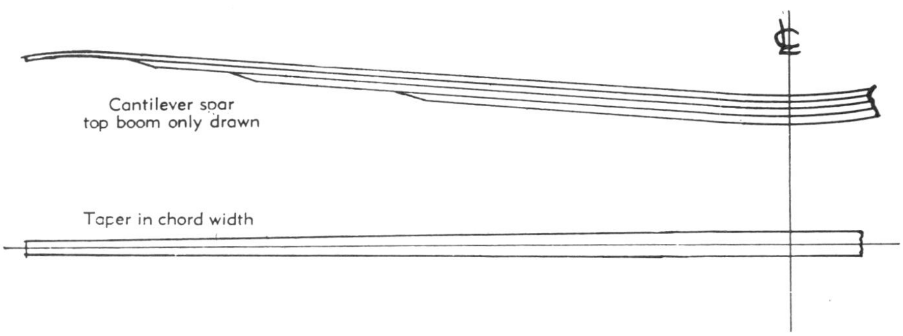

FIG. 4 Some spar booms taper in chord as well as in individual chord depth. Where all ribs are different chords, this variation in spar gap presents no problem.

they also have to conform to the shape of the spar during glueing.

Then comes the job of mating the two booms together with the shear webs, intercostals and blocks in place. If the spar has to be tapered in width as well as in depth (Fig. 4) this has to be catered for as well. This all sounds quite a major job but don't get disillusioned.

There is nothing desperately difficult in building a one-piece cantilever spar!

LAYING OUT THE SPAR

Throughout this article in reference to the spar and spar booms, "top face" refers to the top normal to the spar; "lowermost" or "bottom" face means the lower edge of the spar corresponding to the underside of the wing. This is irrespective of the positions of the booms in the jig. Likewise, "innermost" refers to that part nearest the spar center-line. "Uppermost face" is that uppermost in the jig; "lowermost face" is that against the base of the jig or spar table.

Every spar has a center-line or datum from which the spar is drawn on the plan by the designer or draftsman. This center-line may be a horizontal datum from which the upper and lower boom distances are shown as "plus" dimensions (Fig. 5) or the datum may be the actual center of the spar, expressed as an angle (shown either in degrees or as a vertical distance along a given base length) from which ordinates are given for the booms as "plus" and "minus" dimensions.

Reproduce this line carefully on the spar building table (Fig. 7, page 18, February issue SPORT AVIATION). This is made easier if the table is planed smooth first and painted with an egg-shell or matt white paint.

From this center-line, set off the upper line of the top boom and carefully draw in the positions of all the laminations which go together to make that boom. Repeat this operation for the lower boom. Now check it over carefully for mistakes; any errors should immediately show up as kinks in the gradual curve or straight line of the spar edge.

Should the spar be tapered in chordwise thickness, do not worry about it but assume the spar to be equally thick throughout its length at this stage. If the laminations are, for example, 2½ in. wide by ½ in. thick, order your wood planed to

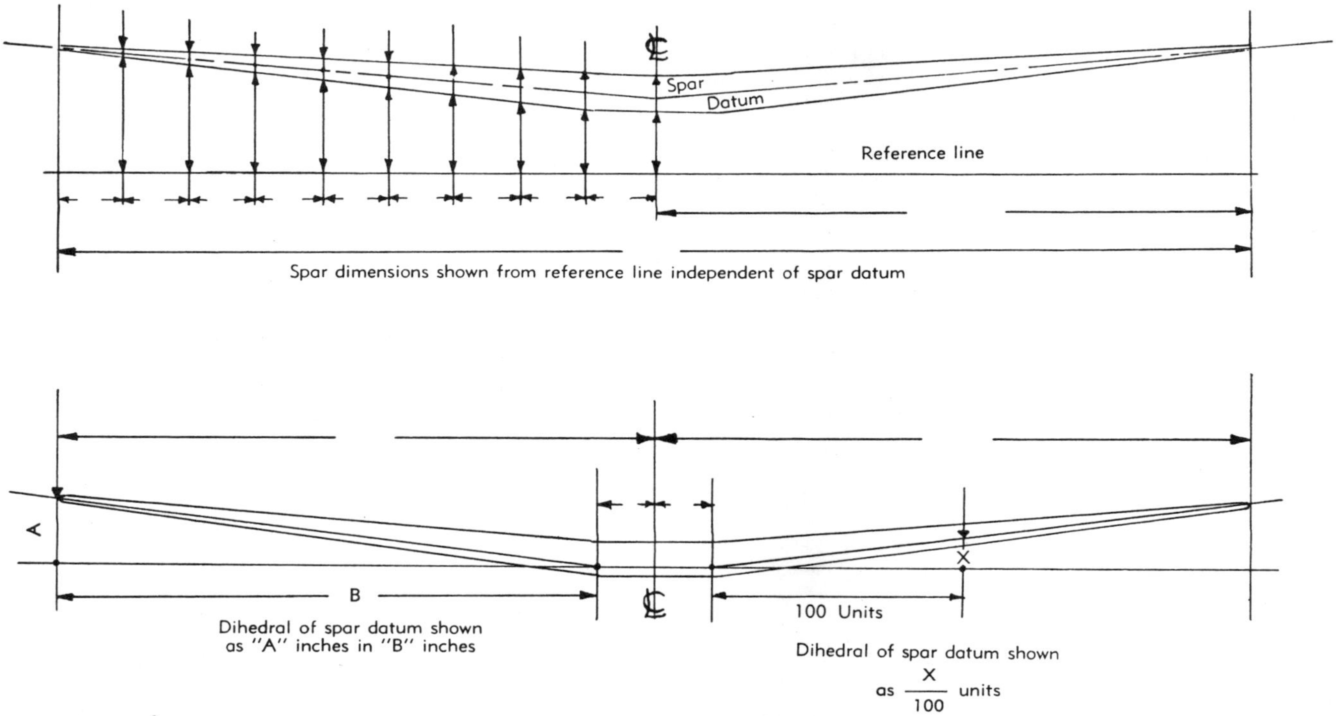

FIG. 5 Methods of dimensioning a cantilever spar. Since it is extremely difficult to draw a spar datum at the correct angle in degrees, one of the above methods is normally adopted.

2⅝ in. wide by ½ in. thick. The added width is to allow for cleaning down to size with the plane after glueing up. Glue and screw a number of heavy wooden blocks to the top line of the upper boom which will be made first. Allow for ½ in. thick loose packing between the blocks and the spar boom. This is so that when the upper and lower booms are glued together with the blocks and intercostals between, pressure can be applied between the two booms after assembly in the loose state with the packing pieces removed. The blocks should be spaced no more than 8 in. apart. This procedure is illustrated in Fig. 6.

You will now need a quantity of large "C" clamps; approximately one per foot run of the spar. These are expensive items and it may well be that you can gather in some from friends, the local garage or timber mill and joiner rather than stand the expense of purchase of a large quantity which may never be used again—until the next airplane comes along, that is.

The objective is to clamp the laminations together, one boom at a time,

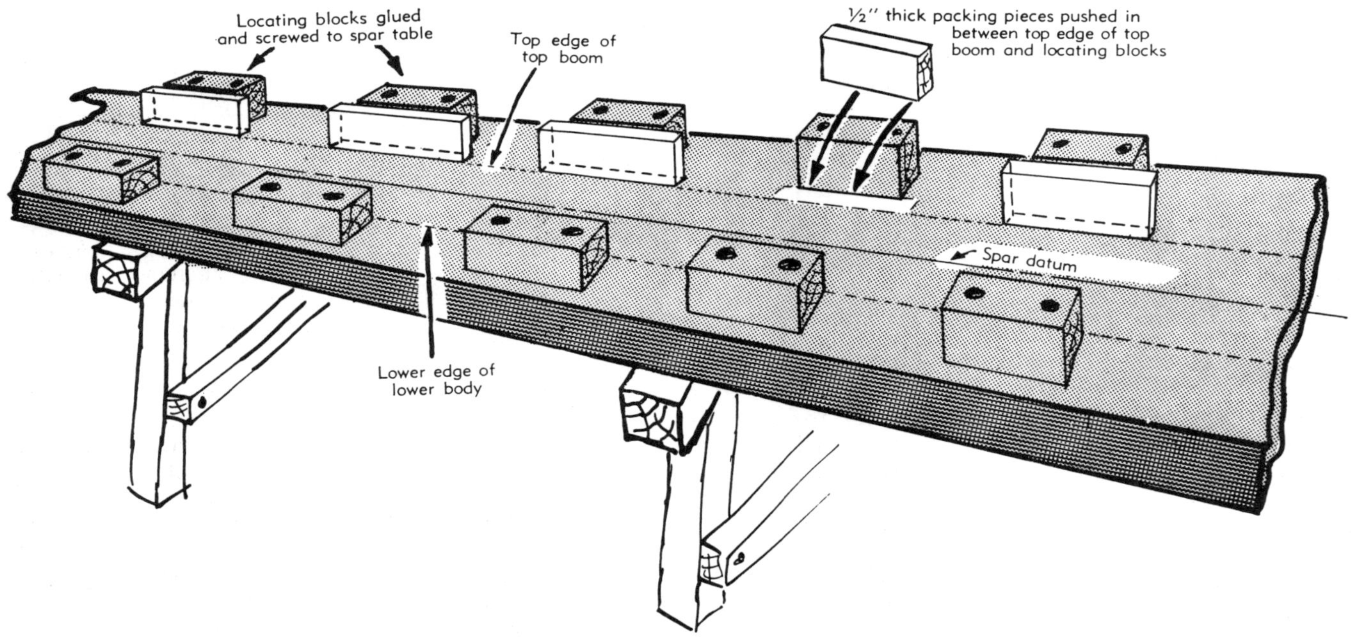

FIG. 6 ½" thick packing is used to facilitate clamping up both booms and the intercostals for glueing. This enables the pieces to be assembled "loose" and insures an even glue line.

FIG. 7 Glueing up laminations. Rub each one to the previous one to exclude air.

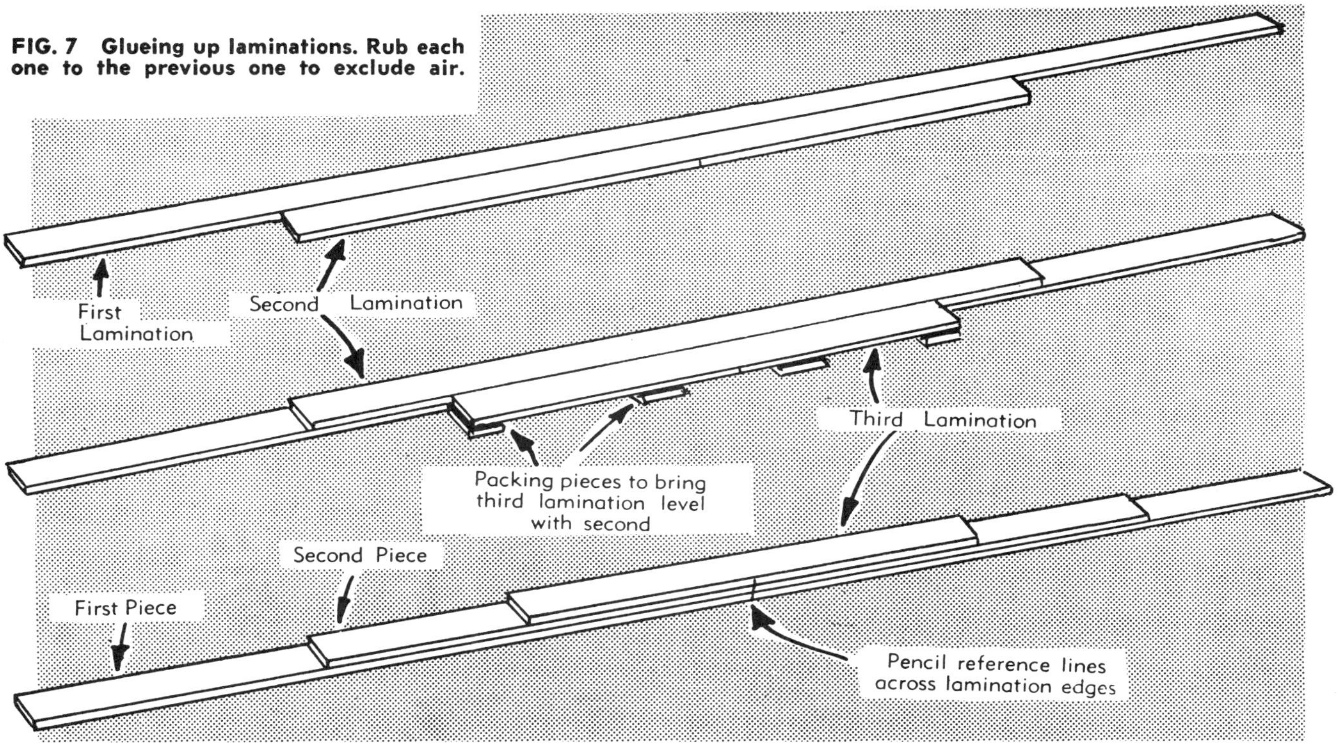

and at the same time to clamp them to the jig blocks so that they conform to the shape of the spar top and bottom profile.

Prepare all the laminations, insuring first, of course, that the timber is Grade "A" spruce, free from disease and defect. If any lengths have to be scarf-jointed, do this now and clean them up on all four sides before assembly. Feather the edges of any laminations which diminish (Fig. 2).

Make one boom at a time—there will be insufficient clearance in the center of the spar jig to permit opposing rows of caul blocks and clamps. Start, therefore, with the top boom and set all the laminations in place. See that the ½ in. thick packing, described earlier, is in place at each jig block.

Using hardwood caul blocks of ample size and with their ends radiused, clamp the laminations to the jig blocks. This assembly is a dry "dummy run" to see that all is ready before starting glueing. Draw a pencil reference line across the edges of the laminations at the center of the spar to aid re-assembly.

Unclamp and remove the laminations. Leave the clamps in their correct positions and set to the approximate gap to save fiddling when the glueing is being done.

ASSEMBLY OF THE BOOMS

Mix up an ample quantity of glue. With modern synthetic resin adhesives and good clamps it is better to overglue and rely on squeezing out a lot than to risk an improper, irregular glue layer through cautious spreading.

Set the outermost lamination (the top one) upside down on a level bench or clean floor. Set the second lamination the right side up next to it and in its correct position to the spar center-line (Fig. 7). Evenly paint glue on to this second lamination and the correct area of the top lamination. Invert the second lamination on to the top one and rub the two together to exclude all air. Do not worry if the glue squeezes out past the end of the second lamination — this is easily cleaned off with a rag before clamping up and further excess again after clamping. Slide the second lamination along until the reference marks on the two laminations coincide.

Now set the third lamination alongside the previous two and pack it up level with scrap wood to aid glueing. Glue and turn it over on top of the other two, rub the pieces together, align the reference marks and proceed in the same manner with any further pieces. Always glue both surfaces; always rub together the strips as a very large glue surface area is being dealt with. If it is necessary to separate two pieces after this on re-glueing, apply more glue to one of the surfaces and proceed as before. On the subject of separating pieces after glueing and rubbing while the glue is still wet, because air is excluded from the wet joint, to try to pull the pieces apart can lead to damage to the wood or even breakage. The only sure way to separate is to fan the pieces, thereby shearing the glue line until the pieces may readily be pulled apart.

With the help of an assistant, carefully lift the glued mass into the jig which will have first been sealed against glue by being painted with hot paraffin or by being lined with thin plastic sheet.

Check that the laminations have not twisted or fanned and, starting always at the center or center-section part of the spar, begin clamping up. Work outboard one clamp at a time each side all the time watching to see that as pressure is applied the pieces do not fan or twist. Again, make certain that the clamp is placed so that pressure is applied along the center of the pieces and not to one side (Fig. 8).

There are a number of ways of getting the correct setting pressure but with modern synthetic resins, tolerances are very much greater than with casein glue or earlier resins. If the glue is spread evenly (as, for example, by means of a roller or squeegee) an even pressure can be obtained by endeavoring to squeeze out an even bead of glue along each joint. It is stressed that this denotes

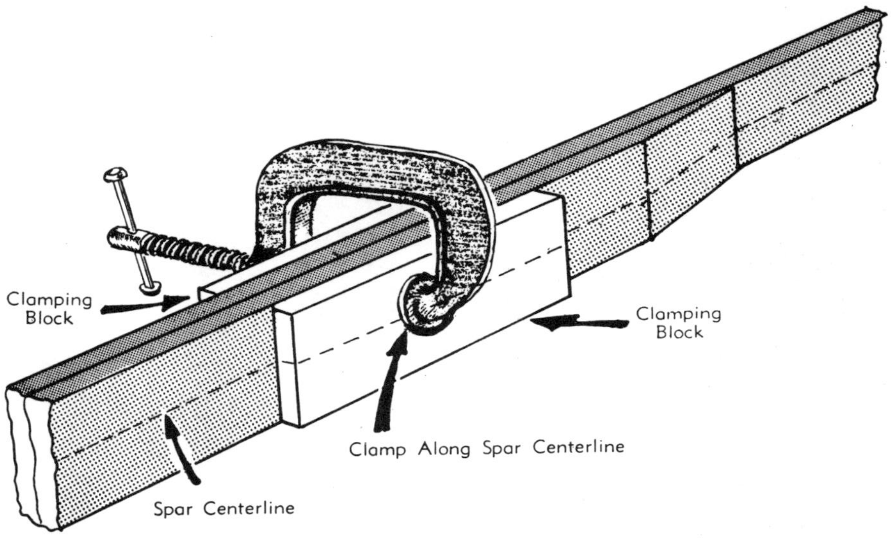

FIG. 8 Position all clamps along the centerline of laminations to insure even pressure and reduce the chances of fanning.

only even pressure and not how much pressure.

With some proprietary glues, the manufacturers supply a chart of recommended setting pressures which must be followed carefully.

Once the correct pressure has been established **using a set size of clamp** each clamp may be tensioned using a torque wrench.

ACCELERATED DRYING

Because a spar boom contains a lot of wood and therefore a lot of glue, normal setting and clamping times at room temperature should be at least doubled. If accelerated glue setting techniques are to be used, rig the elements individually to suit the thickness and number of laminations over a constant area. For example, it is pointless (and unwise) to use a continuous double strip element around the top and bottom surfaces of the stack of laminations because the varying thicknesses of the spar will result in curing at different times. While, say, two hours may be needed at a given setting for the thicker portions, only 20 minutes may suffice for the thinnest part.

By separate wiring, the elements over the thinner parts which set more rapidly may be switched off at the end of curing. Avoid excessive gaps in the element sections. No more than ¼ in. should separate the different circuits and always take the element over the feathered edge from the "high" side (Fig. 9). The circuit between top and bottom sides is coupled with bridging loops **of the same element strip.** Attempts to use

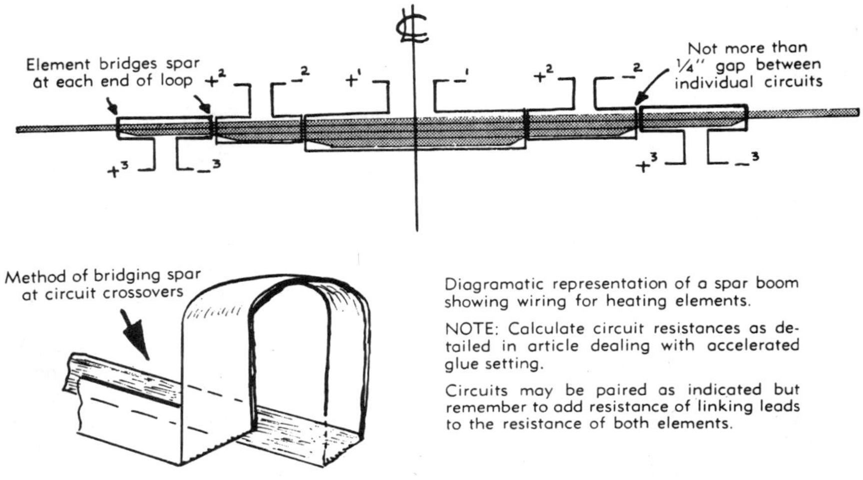

FIG. 9 Accelerated setting of a laminated spar of varying thicknesses using loop elements.

wire could lead to excessive local heat and scorching due to incorrect resistance or arcing.

There must be no doubt that the boom has properly cured. Under normal setting conditions, leave the boom clamped for double the time stipulated on the glue pack; with accelerated heating by means of electric elements clamped in the jig- exceed the normal heating times by 10 percent and then leave the boom for a further 100 percent of the total time.

When the boom is set, unclamp it and clean off the excess glue on the upper most surface using an old plane. At this stage, do not go beyond cleaning down to the bare wood.

Lift the spar boom out of the jig and set it aside in a clean, dry place. Now repeat the whole procedure for the lower boom.

TAPERING CHORD WIDTH

Some aircraft employ cantilever spars which taper in chord width as well as in depth. In the Minicab, for example, there are but a few laminations to the spar, the diminishing loads towards the tips being taken into consideration by reducing the chord width towards the tip. This saves the spar from being excessively heavy as well as making for a true cantilever using the minimum number of laminations yet at the same time reducing the cross-sectional area outwards to the tip.

Scribe a dead true centerline along the top surface of the top boom and the bottom surface of the lower boom. This can either be done step by step with a small rule, pencil and straightedge, or with a pencil mounted in a traverse and scribing block run along either the spar table or the uppermost face of the spar.

Now mark on this centerline the position of each rib and, from the centerline at each rib station, set off the chord widths to each side, not forgetting to allow for the thickness of the spar ribs. Join up all these points with pencil and straightedge and re-check them carefully.

Put both booms back into the jig and pack them outwards against the jig blocks with scrap wood and wedges. With a jack plane remove the boom material down to the tapered chord width line on the uppermost surface of both booms. Work from the centerline outwards so that you are planing with the grain and not against it. The plane iron will need frequent resharpening because you are cutting through the thin but

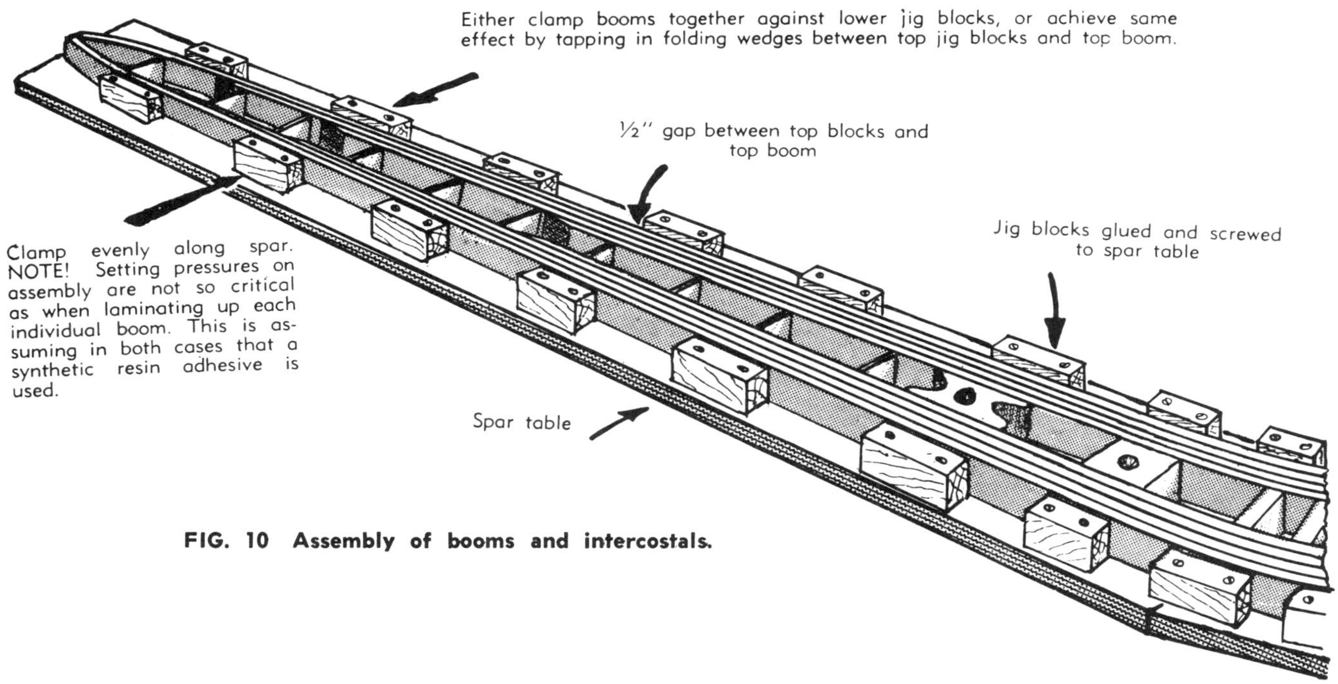

FIG. 10 Assembly of booms and intercostals.

nevertheless hard glue lines as well as the wood.

When the correct amount has been taken off, prepare to glue in all blocks, intercostals, etc., and also the ply shear web on the uppermost side. The reverse or underside of the spar is not tapered until this has been done and the spar locked true.

Remove the ½ in. packing pieces from between the upper boom locating blocks and the boom itself. Glue in all spar blocks and verticals, carefully rubbing each joint and insuring an even fillet of glue. These vertical members are held in place by small tack blocks on each side set in the spar table. This insures that they are at the correct angle to the spar datum; a check with a small try-square is then necessary to bring the piece vertical to the uppermost spar face or datum as the case may be.

If spar and jig have been accurately constructed, it should only just be possible to re-insert the ½ in. packing pieces between the top boom and the top locating blocks. However, these need not be replaced as the object now is to clamp, via suitable caul blocks on the spar upper boom, through the spar to the lower boom locating blocks. This is shown in Fig. 10. Do not allow the spar booms to move relative to one another during clamping.

As soon as the glue has set, unclamp. The spar will remain tight in the jig. With great care, plane and sand the joints between the booms and the blocks, verticals, etc., so that the uppermost surface is presented smooth for the ply shear web.

Since it is impossible to obtain plywood sheets much over 8 ft. in length, strips of ply must be scarfed together to make the web. As with the jointing of individual boom laminations, arrange to stagger the scarfs so that, when the second web is applied to the lowermost side of the spar, the joints will be staggered by half a pitch.

The ply strips for the web should be at least two inches wider than is actually required to permit shuffling during glueing. Sand smooth the scarf joints between the strips. Now evenly sand the entire length of the surface which is to be glued. Use a paper of a grade between medium and fine and wrap it around a scrap block of 2 in. by 4 in. timber.

Because it is impossible, without excessive trouble and effort, to mark out the areas to be glued on the ply, resort to the following trick. Apply a coat of glue to the spar booms and all members in the spar which will receive the shear web. Lay the web immediately on top and press the ply on to the booms throughout the spar length. Now peel off the ply. A "print" of the glue areas will be revealed—in the shape of a thin glue coat. It is now simply a case of applying more glue via a brush to the spar and to the ply.

If possible, "roll" the ply strip on to the booms. This helps exclude air from the large surface areas of glue. Now take a roller squeegee and roll over the top of the ply, forcing it into contact with the spar beneath.

Tack along all booms and members using aircraft brads in either cement-coated steel or brass. To tack up the spar completely requires several helping hands and you should instruct the wife to lay off the coffee and biscuits until the last tack is driven. You must tack to a pattern, starting equally from the center and working outboard one rib bay at a time on both booms. This way, the ply will go on smoothly. Failure to follow this could result in "oil canning" or, even worse, a fold of surplus ply necessitating the complete removal of the ply and a re-start.

Allow the spar ample time to set both with normal setting and accelerated curing times.

Lift out the spar and plane off the flash of the ply web back to the booms. Turn the spar over so that the open side is now the uppermost face. If the spar has been tapered in chord, block it up evenly along its length so that the centerline is parallel to the working surface. It may well be that the application of one ply web will have caused the spar to bow along its length. This is not important as the application of the second web will counteract this. But it is vital that the spar be reset level using clamps and screwed blocks.

BUILDING SPARS . . .

Repeat the procedure for tapering the spar booms, this time taking care not to split the intercostals which also must be diminished.

At this stage, the inside of the spar must be painted or varnished. Keep the preservative clear of the faces to which the second web will shortly be glued.

After the varnish has dried, prepare the glue for the final ply web which has already been scarfed up as before. Using the trick of printing the glue areas in glue, mark the inside of the web. Peel it off and varnish the bare areas. Do not let varnish contaminate the glue areas and paint to within 1/16th of an inch of these areas.

Apply more glue and affix the web as before.

Upon the setting of the glue, remove the spar and plane off the second web flashes. The spar is now complete. If it is necessary to profile the spar to suit the rib camber, this can be done using an adjustable square and the smoothing plane.

There is one more vital operation to perform before nailing on the ribs. This is to mark out the locations of all bolt holes for fittings and so forth. Establish first which is the front face of the spar and mark off on this. Drill all the holes, wherever possible opening them out to full-size. This is because when the ribs are on, it is very difficult to sight accurately for a drill and one has to resort to making up extension drills to get into places which have suddenly become awkward.

Do not forget the basic point in drilling holes through wood; clamp a block of scrap wood behind so that the drill penetrates into it. This prevents ragged and torn wood fibers where the drill breaks through.

Remove all sharp edges with fine sandpaper. The spar is now complete and ready for the assembly of the cantilever wing.

When all ribs and nose ribs have been glued and tacked into place, profile the leading edge portions very carefully. The wing being tapered, any irregularities in the nose portion of the ribs will result in badly buckled ply covering. If the wing sweeps sharply, it will be necessary to pare away at each individual capstrip to bring them to the correct angle. If the change in forward chord is not great, this need not be taken into consideration.

Completion of the wing is now quite straight-forward save that the wing is in one complete piece as distinct from two panels. It is advisable to check the incidence of both sides before ply covering the leading edge as, with a one-piece wing, any discrepancy in incidence due, perhaps, to trestles or incorrect rear-spar or false-spar packing will be locked in.

In a future article will be detailed a production technique for the construction of cantilever spar booms.

The one piece wing of the all wood Emeraude represents many hours of construction time. Note front and rear spars which have to be constructed and jigged so wing has required twist at the tips.

The Ultimate Glue

By Michael C. Myal, EAA 7978

I'M SURE IT has been your experience, as it has been mine, when asking about glue to receive some of these typical answers: "CAM 18 specifies Resorcinol"; "Use MIL Spec glues, if the number is on the can, it's OK"; "I've used plastic resin and it is completely waterproof, and cheap too"; "Don't use epoxy, it has peel strength." These well-meaning statements, however misleading, convinced me to decide on using Resorcinol glue. I was sincerely influenced by well-meaning people to arrive at a decision which I later determined was second best in terms of strength, effort, cost, and quality.

The builder of a wood aircraft or wood aircraft components was always faced with the same baffling problem . . . which glue from a bewildering assortment of products should he choose for his project. This article is intended to provide the builder with the information by which he can better assess the merits of available glues and arrive at a positive conclusion based on fact. This paper also discusses the experiences of the writer and others in Detroit, Mich. Chapter 13 with a special formulation of epoxy glue which, in practice, and in our opinion has been found to be superior to other known types of wood adhesives for aircraft structures.

HISTORY OF AIRCRAFT GLUES

The story of wood adhesives is an exciting parallel to the history of technological progress. It began with the use of by-products of organic matter and has now reached the advanced stage; the development of carefully tailored synthetic compounds for each stringent requirement. The first natural glues, obtained from the bones and hides of animals, were used for little more than sealers for mechanical joints incorporating dowels, dovetails, and tenons. It was the infant aircraft industry of the 1900s which first recognized the possibilities of structural joints.

Blood albumin and casein glues were developed especially to solve the weathering problem and the high strength requirements of the airplane. Manufactured from organic matter, these glues suffered from eventual decomposition, and some successful attempts were made to introduce additives to prevent this deterioration. The discovery in the 1900s of Bakelite, a phenolic thermosetting plastic, led to the development of similar synthetic materials for use in bonding plywood laminations. These early phenol-formaldehyde resins required heat and high pressure during the cure cycle, but did finally achieve the long sought after waterproof and chemically inert structural bond. The need for a companion glue that could be used at room temperatures was apparent, and the demand resulted in the eventual formulation of Resorcinol resins in the late 1930s. Parallel work was responsible for the development of urea-formaldehyde or "plastic resin." Resorcinol was used extensively during World War II, and this wide spread use created for Resorcinol an "image" of strength and stability which persists to this day.

In the post-war period, the rate of development of synthetics was accelerated, producing such diverse materials as dacron, polyester, vinyl, dynel, epoxy, etc. The aircraft industry, hurt by the surplus market and rising labor costs, experimented with plastic structures. Unsuccessful in those attempts, the industry shifted to the extensive use of aluminum alloys, for the most part

FPL No.	Parts Ingredients to 100 of EPON 828	Shear Strength	% Wood Failure
1	(clear) 13 DETA	3133	36
2	(clear) 8 DETA	3098	29
3	(clear)8 DEAP	534	0
4	13 DETA, 10 walnut shell flour	2903	14
5	13 DETA, 10 polysulfide	2905	36
6	13 DETA, 20 polysulfide	2788	40
7	13 DETA, 30 polysulfide	2512	18
8	66 Polyamide	1756	0
9	13 DETA, 10 Kaolin	2149	20
10	13 DETA, 20 Kaolin	2516	20
11	13 DETA, 30 Kaolin	2278	18
12	13 DETA, 40 Kaolin	2264	5
13	13 DETA, 10 glass fibers	2921	1
14	13 DETA, 10 pumice	2736	0
15	13 DETA, 30 T_i0_2	3636	53
16	13 DETA, 30 T_i0_2, 15 Thinner	4102	76
17	13 DETA, 15 T_i0_2, 15 Thinner	3616	71
18	13 DETA, 15 T_i0_2, 15 lacquer	3542	66
19	13 DETA, 15 lacquer	3390	86
20	13 DETA, 15 T_i0_2, 12 acetane	2998	68
21	13 DETA, 12 acetane	2928	51
22	13 DETA, 15 T_i0_2, 12 MIBK	3076	60
23	13 DETA, 12 MIBK	2803	77

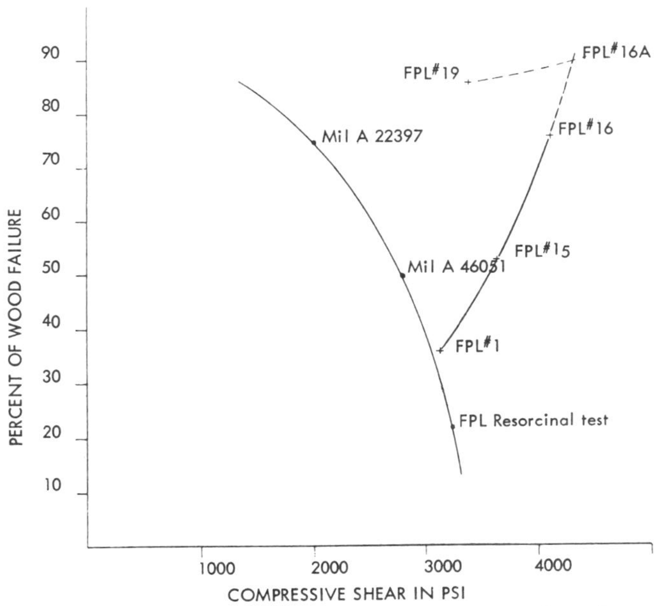

FIG. 1 Excerpt from the FPL report shows various formulations of EPON 828 and their strength values (right). Values are plotted against MIL Spec values for Resorcinol. Note the superiority of FPL 15 and 16. FPL 16A should perform even better due to higher penetration of the wood structure (predicted value).

abandoning wood construction. As a result, competition in wood adhesives ceased, and Resorcinol still remains with us as the basic aircraft wood adhesive.

Recently the aircraft industry has forged ahead in its use of new structural joining techniques. Adhesives, rather than rivets, are used extensively, fastening metal and plastic on such aircraft as the B-58, the Bede BD-1, the Piper "Papoose", the Beechcraft "Musketeer", and others. These adhesives are primarily epoxies specifically formulated for each application.

EPOXY WOOD FORMULATIONS

As early as 1954, the Forest Products Laboratory of the United States Department of Agriculture in Madison, Wis. conducted as a part of its continuing study on adhesives an exhaustive project on epoxy adhesives. During the first phase, the laboratory tested manufactured formulations submitted by various companies for suitable wood joints. These were general purpose epoxies, and not intended for specific wood applications. Tests of these glues showed that these particular formulas were not satisfactory for structural gluing.

Undaunted by these results, the FPL conducted further tests using a base epoxy resin, Shell EPON 828, and a curing agent DETA, diethylene triamine. The shear block tests were run on this resin and curing agent, using various fillers and dilutents (Figure 1). Figure 1 also includes the plots of two MIL Spec minimum strength requirements, and the experimental FPL results on Resorcinol. A solid line connects these three points. It can be assumed that any glue performance plotted above or on the right of this curve exceeds MIL Specs.

In shear blocks tested by the writer, one formula . . . FPL 16 . . . was modified by the addition of three more parts of lacquer thinner (See FPL 16 A on Figure 1) to increase penetration and percent of wood failure. Eighteen hard maple blocks were prepared, six each at contact pressures of 1 psi and 2 psi. The glued surfaces of 3.5 sq. in. were subjected to compressive shear in a hydraulic press and measurements of the load at failure were recorded. Due to the inadequacy of the jigging, these test loads varied from 3,000 to 7,200 lbs.; in each case the block was observed to shift before failure. This short series of tests substantiated the precise results of FPL 16 in the writer's opinion. It can be concluded that this formula is superior to Resorcinol in developing high compressive shear values, essential to structural gluing.

An eyedropper, measuring spoon, and graduated baby bottle can substitute nicely for elaborate weighing equipment. FPL 16A is supplied in metal containers, while the curing agent is packed in glass jars. Cheap brushes of any type can be used to spread the glue.

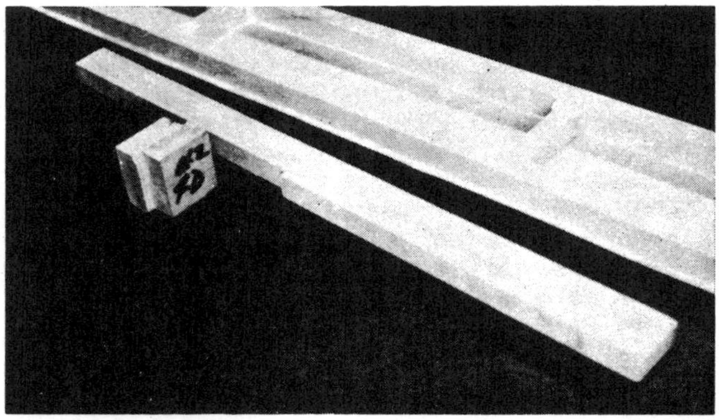

FIG. 3 This sound shear block (foreground) withstood 5,000 lbs. before slipping the test jig. The middle piece is a lamination of eight ¼ in. spruce pieces as trimmed from the author's Minicab main spar. The assembly is a partial view of the rear spar showing web joints.

OTHER ADVANTAGES OF FPL 16 A

We have discussed the superior strength of FPL 16 A as compared to the old standby, Resorcinol. There are other characteristics of FPL 16 A which make it even more attractive for homebuilding. Being a typical epoxy, FPL 16 A possesses gap filling properties. The glue line can vary from one to eleven mills without any appreciable loss in strength, as compared to Resorcinol which is vary sensitive to variations in glue line thickness. Maintaining his workmanship standards constant, the homebuilder using FPL 16 A will automatically improve joint quality over prior glue lines, due to this unique feature.

Another characteristic of FPL 16 A is its long assembly period. No longer does the homebuilder require additional help to hurriedly skin a wing panel, laminate spar flanges, or glue large and complicated areas. Open assembly time is increased from 10 minutes for Resorcinol to about 1½ hours for FPA 16 A at room temperatures. There is ample time now to align pieces, have that cup of coffee, and actually enjoy the entire process!

What about help for nailing down tack strips? FPL 16 A comes through again for the homebuilder! Contact or lower order pressures are sufficient to obtain optimum glue line strengths.

Tests by the Forest Products Laboratory indicated that high clamping pressures and thin glue lines produced weaker joints than those resulting from low pressures. In skinning a turtledeck for my Minicab with 1/32 in. aircraft plywood over ¼ in. mahogany formers, I formerly used a common office stapler to locate and hold down the plywood. Straps could have served the same purposes. FPL 16 A virtually eliminates the tedious clamping and nailing of wood structures, saving effort, and stronger structures will result from the absence of damaging nail holes.

There are other adventures of epoxy FPL 16 A over Resorcinol. The resin is white and non-staining. The glue line does not shrink or produce by-products during cure. The odor is pleasant, and the wife will love you for it. The resin has a long shelf life, and does not age once the can is opened. The glue produces a fillet which has the effect of lowering stress concentration at joint intersections. The glue spreads thinner, effecting a cost saving of 50 percent over other glue.

USING FPL 16 A

This glue consists of two components, both liquid. The resin component consists of 100 parts (by weight) of Shell EPON 828 resin, 30 parts titanium dioxide, and 18 parts lacquer thinner. (The lacquer thinner consists of, by volume, toluene, 74 parts; butyl acetate, 41 parts; and butyl alcohol, 11.7 parts). The curing agent component is DETA (diethylene Triamine). The mixing ratio is 148 parts resin to 13 parts of DETA by weight, or 10 to 1 by volume. This formula contains a calculated excess of curing agent to insure proper curing. DETA is a drying agent and should not be allowed to contact the skin, and breathing of its vapors should be avoided. A thorough wash up and the use of hand lotion will prevent any skin problems from arising.

Several simple rules should be followed. First, the resin component should be thoroughly stirred, as some separation of thinner and white filler will occur with time. A paint paddle is very effective for this purpose. Second, quantities of stirred resin and curing agent should be carefully measured. Third, thorough mixing is essential.

Care should be exercised in preparing small quantities by the volume method. For example, in making small batches of one to three tablespoons, I found that calibration of my measuring spoons was essential. For every measure of resin used, I added the required 10 percent of DETA with an eyedropper.

All of my glue mixing follows the volume method, with excellent results. For large quantities a baby bottle is used as a graduated flask to calibrate glue pots with water. The level is marked, the container dried, and the resin poured to the mark. Another smaller baby bottle is used to establish the necessary 10 percent of curing agent. When ready for gluing, I add the curing agent to the resin and stir thoroughly. MOST IMPORTANT ARE THESE SIMPLE MIXING STEPS:

1. STIR THE RESIN COMPONENT THOROUGHLY.
2. MEASURE OUT COMPONENTS CAREFULLY, PARTICULARLY FOR SMALL BATCHES.
3. MIX GLUE THOROUGHLY.

The glue, in use, resembles somewhat thinned out four hour white decorative enamel in consistency and color.

Economical dime store brushes or acid brushes for small areas can be used to spread the glue in an effective manner. One quart of mixed glue applied to both surfaces will join approximately 35 sq. ft. of Sitka spruce,

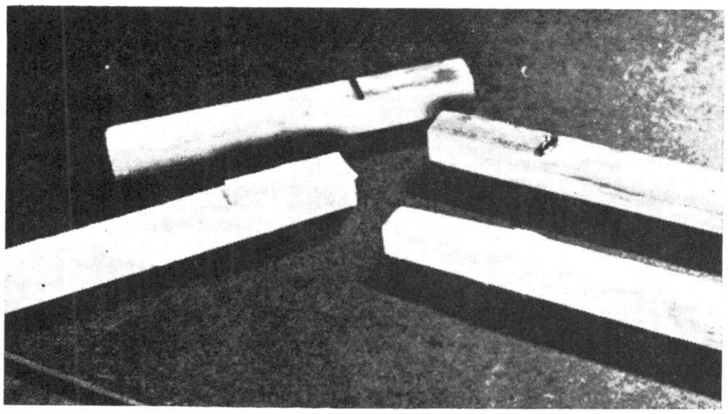

Resorcinol lap joint tension test (rear) shows glue line failure while FPL joint pulls wood, was devised to check supposed "low peel strength" characteristics of epoxy. ANC-18 and 19 do not discuss "peeling", since such wood joints and loadings are not aircraft practice.

or 140 sq. ft./gal. According to the coverage printed on the Resorcinol label, a gallon will cover approximately 90 sq. ft. (both sides). FPL 16 A will cover 50 percent more area than Resorcinol.

The glue is painted on both properly prepared surfaces, and the pieces are mated in a sliding motion to exclude air. No pressure is necessary during the curing cycle. To insure intimate contact of plywood skins or thin gusset plates which tend to flex or curl, clamps can be used at the extreme corners of the joints, as well as nails, staples, or weights. The applied pressure should not squeeze out the glue; only the amount that provides positive contact is necessary. On large areas of skin, a staple gun of either office or compression style can be used to advantage to hold plywood to the ribs or corners during the cure. Nails, for the most part, can be eliminated from this gluing system which is a distinct advantage for the rib builder. Curved or formed laminated assemblies will require clamping devices but, again, the pressure required is only that necessary to maintain surface contact.

The pot life of FPL 16 A glue depends on the room temperature, the quantity of glue, and the rate of usage. Two of us applied a quart of glue at 70 degrees without any appreciable change in consistency, although at the end of a half hour period, the emptied container was noticeably warm. It is suggested that glue requirements larger than quart size be pre-measured and mixed as needed. A quantity the size of a baby food jar, used intermittently, will last about one hour before the consistency is affected. If the glue is brushable, it will produce an effective joint; any batches which become difficult to apply with the brush should be discarded. Lacquer thinner is used for cleanup and, as a rule, brushes can be re-used several times.

At room temperature, this glue will produce a joint overnight which has sufficient strength for handling. A full strength cure depends on time and temperatures, as with Resorcinol, but a 48 hour period will produce joints which, when tested, pull wood at the glue line. A one week cure at room temperature should be ample to attain full structural strength. At higher ambient temperatures, the glue pot can be submerged in cool water to retard the reaction.

Lower gluing temperatures can be tolerated by the resin, but again the time required to attain strong joints will be extended considerably. FPL 16 A applied at near freezing levels appeared to produce satisfactory joints in brief tests after a post cure at room temperatures for 24 hours. It is recommended that the builder conduct

COMPARISON TABLE

Characteristics	Epoxy FPL 16A	Resorcinol	Plastic Resin
Strength	x	x	?
Waterproof	x	x	?
Permanence	x	x	?
Gap Filling	x		
Filleting	x		
Working Life	x		x
Shelf Life	x	x	x
Contact Pressure	x		
Mixing and Application	x	x	x
Workmanship	x		x
Efficiency	100%	50%	40%

This table of comparison illustrates the overall efficiency of FPL 16 A as compared to commonly used Resorcinol and plastic resin.

such tests as necessary to prove to himself the strength of joints assembled at below 50 degrees.

SUMMARY

FPL 16 A is superior to Resorcinol in most respects, including the factors of strength, water resistance, chemical inertness, shelf and pot life, ease of application, joint quality, and cost. In the writer's opinion, this formula will establish a new standard of performance for aircraft wood glues as other homebuilders try and discover its properties.

Most reputable glue manufacturers can prepare this formula, as the ingredients are readily procurable. If you have difficulties in locating a company willing to serve your needs in your area, our local manufacturer . . . Hughes Glue Co. of 277 Leib St., Detroit, Mich. 48207 . . . will custom mix this formula for a very reasonable price.

It should be mentioned that general purpose epoxies are not recommended for aircraft use unless conclusively tested. There are literally thousands of formulas available from reputable companies which work wonders on specific materials, but may be disasterous for joining wood. Some epoxies such as 3M 2216 BA and Shell 907 are satisfactory, but at $30 to $50 a gallon, the arguments for their use become rather weak.

This formula is being used by the author in building his Minicab. Other builders within Chapter 13 have used it and reported complete satisfaction.

The shear block results were shown to our local FAA inspectors and were demonstrated to their complete satisfaction. It should be pointed out that the FAA inspector cannot and will not assume any position with regard to type of brand of glue. He is interested in the strength and waterproof capabilities of the joint and its quality. After samples of your glue joints are made available for destructive testing, they will provide the proof of its superiority.

(NOTE: The reference work "Epoxy-Resin Adhesives for Gluing Wood" by Olson and Blomquist in the February, 1962 Forest Products Journal provided much of the background for this article).

Wood tail surfaces nearing completion. Stabilizer has yet to be covered with plywood. Note wing rib jig in center. Usually just one jig is necessary for ribs with wings of constant chord. Cantilever wings, however, require many jigs as each rib changes in shape and size from tip to root. Only two ribs — left and right panels — will be identical.

Above — Just a little bigger and using stronger wood — but other than that this fuselage goes together very similar to the model airplanes most of us built as youths. Below — Laminated wing tip bow. Thin wood strips are usually steamed then fitted in a jig for glueing strips together.

Epoxy Resins and Private Built Aircraft

By V. M. Draper, EAA 1778

There has been considerable talk and writing about the epoxy resins, how good they are and the uses to which they may be put. They have proved to be superior in many ways to the more common and less expensive polyesters but so far as I know, no one has used them to any great extent in homebuilt planes.

That they are of benefit for aircraft use is proved through their use in bonding the skin on the Convair B-58 "Hustler", although I don't know just to what extent epoxy was used in the Hustler. It should also be emphasized that special procedures were used in its application. However, several concerns are marketing epoxy glues in various forms that are no more complicated to use than Elmer's Waterproof Glue. In every instance the new epoxy glues involve mixing the resin and hardener, with the latter being mixed in quantities from 25 to 100 per cent of the resin.

I have been experimenting with a product called Fuller's Resiweld, which is resin and hardener mixed 50-50, and have obtained excellent results with the utmost simplicity in application procedure. In fact, I am so impressed that I intend to use Resiweld to apply the plywood skin on my Cougar's wings. I have one panel covered on the top with Weldwood plastic resin glue as the adhesive and found that unless one has a lot of help, under ideal conditions, it is hard to get a good job. You have to make good solid joints with adequate pressure or you don't get a good bond. Furthermore, unless you have adequate heat you may not get a dependable job. Applying adequate, even pressure is the most difficult problem facing the users of plastic resin glues.

In my experiments with the epoxy glues I have found that so long as both surfaces mate well only a nominal amount of pressure is required. In fact, I have achieved excellent glue joints without pressure. By good joints I mean bonds that are stronger than the materials glued together. Spruce capstrip and one-sixteenth mahogany aircraft plywood, for instance. I have made several experimental joints with these materials and in every instance either the plywood or the capstrip has broken, rather than the glue joint. Usually, when test stresses are applied the first ply adjoining the glue joint gives way — the plywood comes away from the joint minus a strip of mahogany where the joint was achieved.

I am satisfied I can get a bond that will meet all safety requirements with greater ease of application than other glues. I have tried the epoxy glue in temperatures in the low forties and still get good bonds if sufficient drying time is allowed. This is both advantageous and disadvantageous — where one must glue under adverse temperature conditions one can get a good bond. On the other hand, at least 24 hours under pressure should be allowed under the best conditions.

I said only nominal pressure need be applied; however, under no circumstances should there be a thick glue line between surfaces glued. Thus sufficient firm pressure should be applied to achieve the desired thin glue line. The main thing is to get the two surfaces to mate perfectly and have thorough epoxy coverage. Oh yes, with epoxy one need coat only one surface with the glue. Another advantage of using epoxy is that there is much strength in the epoxy itself, a trait not present in the other types of glue.

Ideal temperature for epoxy glues would range from 70 to 90 deg. F. and this would give a satisfactory joint in less than 24 hrs. On the other hand, a good bond can still be achieved as low as 40 deg. F. by letting the work set longer. I would not advise using epoxy in making rib trusses due to mixing difficulties with such small quantities. The biggest advantage of using the epoxies lies in projects such as plywood wing covering where one has to get the job set up in a hurry after the glue is applied. Because one is more concerned with even pressure, rather than a specific amount of pressure in pounds, greater latitude is permitted in the application of pressure when epoxy glue is used.

In my case I use web belting to apply the pressure, nailing through the belting on flat surfaces where pressure of the belting is insufficient to obtain a satisfactory joint. Or, that is, this WAS the procedure when using plastic resin glues. When using the epoxy I will merely draw the belting across the ribs tight enough to assure a thin glue line. Nails will still have to be used in the transverse belting along the spar and trailing and leading edges. But, the time and labor saved will amount to several hours in the overall job of covering my wings. Best of all, I will be spared the worry of, "Did I apply enough pressure to get airworthy joints?" With epoxy I know that so long as the surfaces mate and are left long enough to cure, the bond will definitely be airworthy.

Despite the fact that the epoxies will dry in cool temperatures I plan to work in a heated shop and also use heat lamps to speed curing in the shortest time possible. With both plastic resin glues and epoxies, at least three days at normal room temperatures should be allowed for curing in order to gain the best possible job of gluing, although it is not necessary to keep the work under pressure longer than the first 24 hours.

You may ask, "Just what are the specific advantages of using epoxy glue?" The first and main advantage lies in epoxy's property of obtaining good adhesion under adverse conditions. The second lies in the lesser skill required to get a satisfactory bond, and the time saving achieved through less complicated means of pressure application. The third lies in the epoxy's own strength.

The disadvantages are: the epoxies must be very accurately mixed and take somewhat longer to dry under ideal conditions. And, as yet they're expensive. But, in the long run I think they have a wide edge in strength and dependability to make their use well worth while.

Trailing Edges in Wood

By Arthur W. J. G. Ord-Hume, EAA 8579

PIONEER BUILDERS — and they were homebuilders in the truest sense of the term — treated the trailing edges of their wings with surprising scorn. Again, it could have been a surprising stroke of brilliance at the time — times change and with them ideas. The prime purpose of this part of the wing, thought our homebuilding predecessors, was to bring the chord to an end and provide something to hook the canvas to. The answer was a piece of wire. The resultant wing presented an attractive scalloped rear edge around which the cloth was stretched in not so attractive wrinkles.

A wire trailing edge is simple, fairly effective and light in weight. It does, however, depend to a far greater extent than any other type of member on the resistance of the rib to buckling and transmitted loads. As can readily be appreciated, a knock on the trailing edge wire can be transferred to all the ribs and the ham-fisted mechanic who clouts the hangar door with your wing can probably ruin the entire panel instead of just goofing up a foot or two of trailing edge wood and a rib.

Again, the wire trailing edge has no inherent resistance to bending and can deflect downwards — or upwards — under the tension of the fabric, producing a highly unaerodynamic effect which will make itself apparent on anything intended to fly faster than 40 mph.

For the same reasons, although not so liable to their effects, the thin tubular metal trailing edge, fixed to the wooden ribs with little wrapper plates, is also to be frowned upon. It is light, simple and effective — but it warps, distorts, buckles and, once dented, proudly proclaims its condition regardless of any action other than Divine Providence or a new section of member.

Why use wood? To begin with, it can be glued to the wood ribs and at once forms part of a homogenous structure. It is easy and positive to work, is capable of simple, effective and quick repair and is far, far stronger than any other equivalent trailing edge for a fabric-covered wing.

To paraphrase this, the old C.3 Aeronca had ¼ in. square capstrips and the trailing edge comprised a flanged "V" section of light alloy which was nailed to the ribs and bound with tape. The whole issue gradually worked loose and one unconsciously found oneself relying on the cotton fabric to keep the wing together.

We can consider four basic types of wooden trailing edge member and each is depicted in the accompanying illustrations. Needless to say, there can be variants and mutations on each of them. Figs. 1 and 2 can be thought of as mainly suitable only for sailplanes and gliders, combining extreme lightness with ample strength. Figs. 3 and 4 are suited to the average homebuilt airplane.

Let's take an analytical look at each one.

Fig. 1 comprises a thin spruce strip, suitably shaped to the streamlined form dictated by the rib. In itself, this strip has little inherent strength, being perhaps only ⅜ in. by ½ in. However, by slitting it on a power saw and inserting a long-grained plywood strip into the resultant slot, the effective strength of the little spruce strip is vastly improved. Furthermore, the ply strip provides an

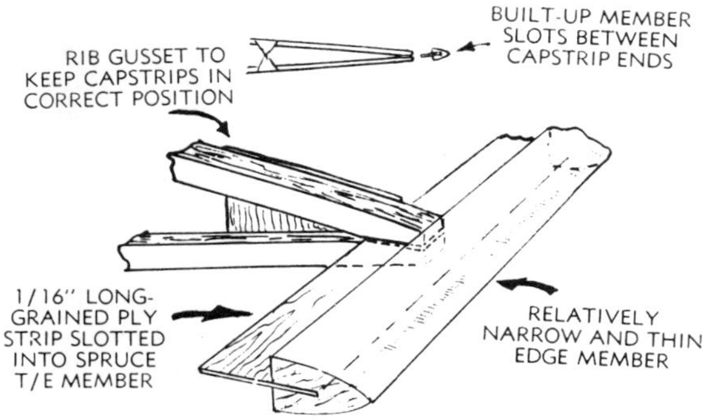

FIG. 1

adequate method of attaching the member to the ribs and, as it forms its own gusset, no additional plywood reinforcement of the joint is needed. The ply strip should be inset into the actual trailing edge member by about ¼ in. and should protrude between 1 in. and 1¼ in., depending on the rib spacing and the actual size of the spruce member. Normal 12 in. rib spacing dictates about 1 in. protrusion.

Fig. 2 is an all-spruce variation on the design shown in Fig. 1. This is more tricky to prepare but requires no slitting and consequent glueing in of the ply strip. It forms a stronger joint to the rib but requires one gusset on the undersurface (always gusset underneath unless special rea-

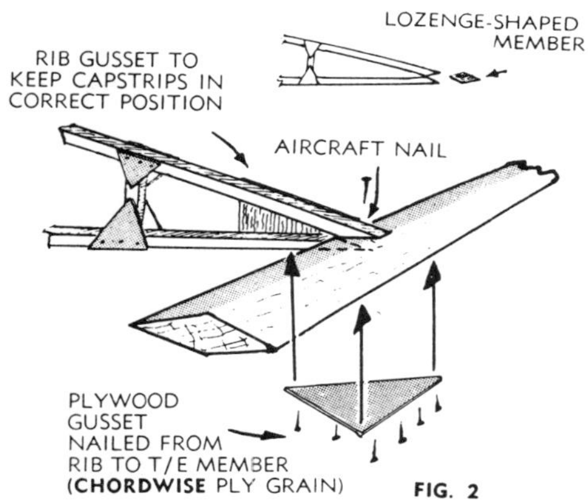

FIG. 2

sons dictate otherwise so as to preserve the top surface) to make a good fixing. This is a light edge member and can be used to advantage on tail surfaces.

Fig. 3 is a very robust member and is designed so as to fit flush with the undersurface of the wing. Attachment is with a gusset. For normal 12 in. rib spacing, a 1 in. or 1¼ in. by ¼ in. member, suitably shaped, is extremely strong. The gusset should be approximately a 3 in. base by 3 in. high triangle with the grain running in the direction of the chord-line of the wing.

Fig. 4 is again very strong and has the merit of needing no gussets. It is ideally suited to the high camber wing where the trailing edge of the rib will accommodate a

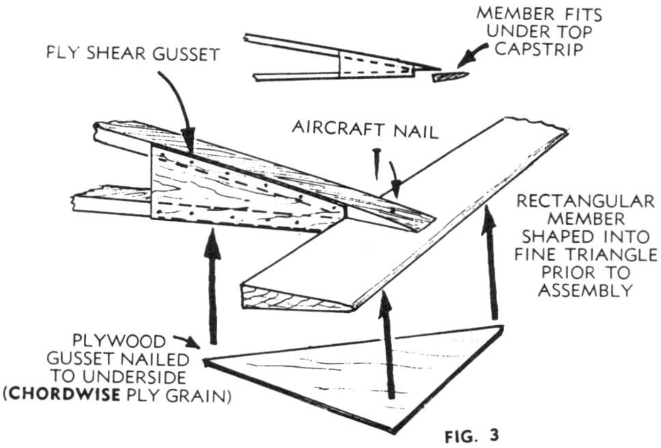

FIG. 3

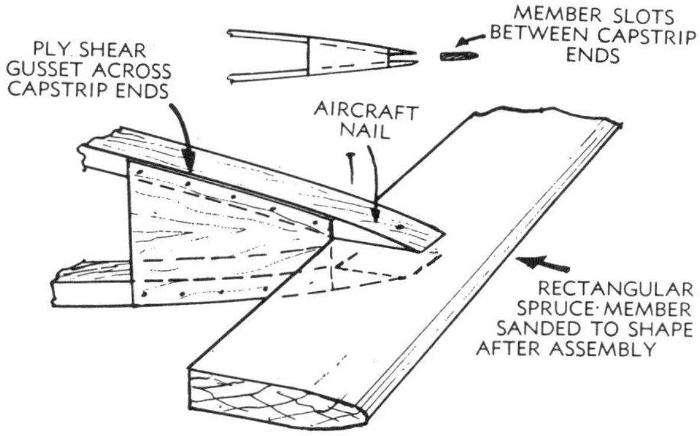

FIG. 4

fairly thick-sectioned member. Conversely, it is unsuitable for a thin wing and should not be used on sections less than about 17 percent. The member can be about 1¼ in. by ¼ in. and suitably shaped to suit the airfoil section.

Why use such heavy sections for the trailing edge of the wing? One of the first things that the "old school" of aero-mechanics was taught was that you never pushed an airplane by the rear edge of the wing. If you de-rigged an airplane, you never stood the wing on its trailing edge. Today, you find just about everybody pushes an airplane by the trailing edge. This is possibly sheer indifference to the commercially-built tin flying machine — or just plain pig-headedness! Quite often the wing leading edge is so cluttered up with pitot pipes, fairings, fences, stall warning indicators, buffet strips, landing lamps, sophisticated cow-catchers and so on that the only edge you can stand the wing on is the fragile trailing edge.

With these tough, yet light members in your wings, you can be sure that no ham-fisted mechanic is going to push your trailing edge flush with the rear spar, no trailing edge standing will make your wing look like a roller coaster.

In conclusion, these four basic design forms can be used equally well on curved wings. Fig. 1 is simply slotted and then glued to a curved strip of ply, the pieces held in a simple jig while the glue sets. Fig. 2 is man enough to be laminated in two strips and so are both Figs. 3 and 4.

There is a saying that you should **build** certain features into an airplane. For the same reason that aspirin is only to ease the discomfort of a common cold — it does not prevent it — I feel that building features into an airplane is only second best. They should be **designed** in right at the start. It is, after all, the details which count. ✲

METHOD OF SPLICING SOLID "I" SPARS

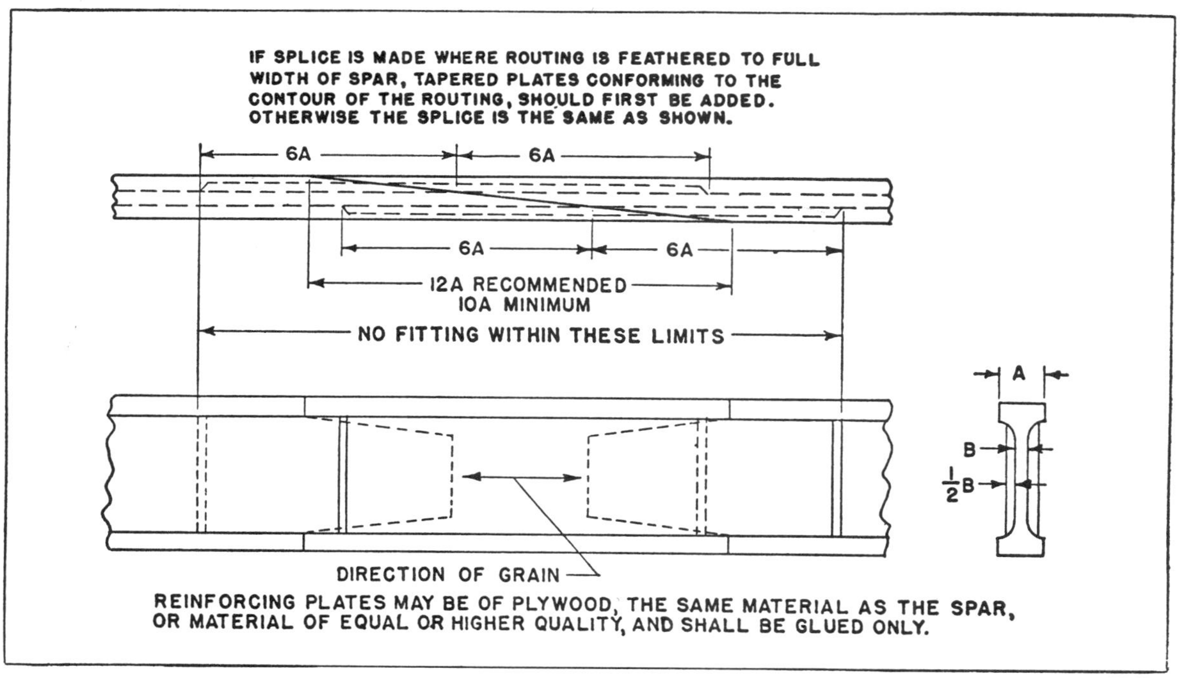

Why Not Build a Wood Aircraft

by Tony Bingelis

WHAT does a wood aircraft look like? It can look like an ordinary production airplane . . . but it doesn't have to.

It can look like a perky little biplane, an amphibian, or maybe like a plush low wing job with a sliding canopy. On the other hand, it might look like something out of the pages of history - a World War I or World War II fighter, perhaps. Then again, it might look like a sporty open cockpit homebuilt obviously intended for fun flying and airport hopping. It can even look like a sleek, speedy, high tech design with a tucked up landing gear, controllable prop and loaded with instruments and avionics. In short, your wood airplane can be constructed to look like any kind of aircraft your mind can conceive . . . but, best of all, it doesn't even have to be painted white like those fast, but expensive, "plastic" ('scuse me, "composite") jobs.

So, why not consider building a wood airplane with a personality of its own . . . one that will not suffer that "look alike" monotony so characteristic of many current aircraft designs.

WHY WOOD?

Well, why not?

Indeed, wood is a natural construction material . . . the only aircraft building material not manufactured by man.

It is readily available, and even the smallest of community lumberyards may carry some of the woods suitable for use in aircraft construction (Douglas Fir, for example).

You can buy a few boards and rip them to size if you have a bench saw. Or if you prefer you can purchase dimensioned aircraft spruce wing rib capstrips and pre-cut longeron materials from one of the homebuilder supply houses.

A builder with a good set of plans in hand, and a strong desire to build, will find that wood is easy to work with and is as strong as most man-made composite materials and light alloys.

Furthermore, an aircraft built mostly of wood will provide you with an airplane for which the overall cost, even today, remains lower than other forms of construction. Actually, compared to metal construction, simple wooden airplanes are cheaper to build and require little or no special tools and equipment other than typical home workshop tools.

Certainly, most of us have more basic experience working with wood than we have with composites and metal alloys.

When you complete your wood aircraft according to the plans, you will be assured of a very stable structure, one remarkably resistant to a broad range of stress factors . . . and best of all it will be fully paid for, easy to maintain, and easy to repair.

A wood airplane, like a composite one, is much quieter than a metal aircraft because wood doesn't "oil can" and resonate noisily like metal.

Perhaps the primary reason wood aircraft are not being built commercially is simply because, by its very nature, wood does not lend itself to automated mass-production. Obviously, this form of construction is not a drawback to the average builder who merely wants to build only one aircraft and can expend the necessary manhours more freely than dollars.

HOW LONG WILL IT LAST?

No construction material will last forever . . . aluminum corrodes, steel rusts, composites decay, wood decays. However, with the proper protective measures you can virtually eliminate rusting, corrosion and decay in whatever material you choose to use.

Wood assembled with good glues and adhesives, and treated with protective coatings, can be made to last for decades.

For that matter, a number of homebuilts built in the early 1930s are still airworthy today. This, in spite of the fact that the glue used in those bygone days was a casein glue.

Casein glue was a dairy by-product. It lacked gap filling qualities and was not waterproof or resistant to mold. That made it far inferior to most glues and epoxies available to builders today.

Even so, poor glue notwithstanding, these wood wonders have not only sur-

You can begin construction of your wood aircraft inexpensively by starting with the wing ribs. Very little expense is involved here and the satisfaction gained with the completion of each rib is hard to convey to non-builders.

vived but a number of them are still airworthy 50 years later.

Today, in addition to the availability and use of superior adhesives and preservatives, the complete aircraft structure is normally protected with two or three coats of clear glossy polyurethane varnish. This varnish is waterproof and does an excellent job of sealing out moisture.

Another protective treatment for wood structures, known as the "West System", is used by many wood boat builders as a means for waterproofing a boat's framing and planking. It is an epoxy that coats and seals the wood quite effectively. Many homebuilders also use it.

Longevity of a wood structure is unlimited provided no moisture is allowed to become entrapped over long periods of time. This means that if moisture enters a wood structure it should readily drain out so that no water remains accumulated anywhere. You can accomplish this with strategically located drain holes. No entrapped moisture - no dry rot.

Incidentally, this same need is applicable to metal aircraft structures, too. These drain holes should be inspected periodically to assure that they remain open.

ARE KITS AVAILABLE FOR WOOD AIRCRAFT DESIGNS?

Yes, but their numbers are not many, nor are these kits as highly prefabricated as those for the current crop of high dollar composite aircraft.

Wood aircraft designs, for the most part, are not available in kit form where the major components, fuselage, wings and tail section are prefabricated. In other words, you will have to assemble these parts from raw materials, or in some cases, from basic wood kits consisting of bulkheads and ribs . . . or more likely consisting of wood, plywood, glue, nails, etc.

Special, hard to make parts like cowlings, canopies, landing gear components, engine mounts and similar parts are available for many of the more popular wood aircraft designs.

However, even if you do elect to purchase some ready made parts, the overall cost of building a wood airplane is quite modest compared to those so-called, complete kit projects because you will be doing most of the construction.

YOU CAN GET STARTED FOR, SAY, $100

As with most projects, even the expensive kit types, building an aircraft begins with the construction of the comparatively inexpensive tail surfaces. This enables you to become familiar with the construction practices, and to develop your skills on the relatively simple and least costly parts of the airplane.

All you need, besides the plans, is some glue, wood strips, a small sheet of plywood and, perhaps, some small aircraft nails. With these basic materials on hand, you can get started on your project by constructing the tail surfaces or, as many builders prefer, building wing ribs.

Should you learn that building a wood aircraft is not for you, you can simply quit - and you will not be out much money at all.

Even so, you will have had some useful hands-on experience and entertained yourself quite inexpensively, too.

If you are a wheeler-dealer type, you might even be able to sell what you have completed to someone else. You can't lose . . . unlike bowling, golfing or attending rock concerts, etc., you will at least have something to show for your modest initial expenditure.

THE THREE EASIEST TO BUILD WOOD AIRCRAFT

I'll bet you have accumulated your share of brochures, visited local builders and once you even bought a set of plans for an airplane that just caught everyone's fancy. It seemed just like the airplane for you until you studied the

The rugged strength of the easy-to-build wood structure of a Boredom Fighter is most evident here.

A wood fuselage ready for its fabric covering reveals the simple method of construction typical of almost all wood aircraft . . . four longerons, truss braces, and stringers. All joints are strengthened with plywood gussets or plywood skins.

The Fly Baby in its element. This is what they mean by grassroots flying . . . it will never get better than this.

plans and concluded that it would be too expensive to build . . . even if you could figure out how to build it.

Well, there is new hope for you to realize your ambition to build and fly your own homebuilt.

Here, for your consideration, are three excellent wood aircraft designs I have always like because I believe them to be:

1. Fun to fly.
2. Easy and inexpensive to build.
3. Enjoying a timeless popularity.
4. Economical to operate (engines are under 100 hp).
5. Safe and proven (many hundreds, actually, thousands have already been built and flown).

These three classics alphabetically are:

1. The Peter M. Bowers "Fly Baby"
2. The Bernard H. Pietenpol "Pietenpol Air Camper"
3. The Bud Evans "Volksplane" (VP-1 and VP-2)

THE FLY BABY

Construction is all wood with fabric covering for this low wing single seater first built in the 1960s. It can comfortably accommodate large pilots. The designer is well over 6 feet tall himself - that should tell you something about cockpit size.

The plans are unique in that they are in the form of a step-by-step pictorial presentation incorporated in a single loose leaf manual. The manual consists of many detailed drawings of individual parts, often shown in full size or isometrically in support of the clearly written instructions.

The manual is very complete and includes a very useful "Bill of Materials" for the major components.

The fuselage is a simple box structure consisting of four longerons and plywood skins. The wing ribs are all the same size. They are cut out of plywood and are reinforced with grooved upper and lower capstrips.

There is even an option for converting the basic low wing Fly Baby to a biplane using the same fuselage, tail and landing gear.

After some 30 years or so, the Fly Baby is a well proven design. Hundreds have been built and continue to be built by builders interested in having a good safe economical airplane to fly.

The construction manual is a pleasure to use and it has proved to be a reliable best friend for many a first time builder.

THE PIETENPOL AIR CAMPER

This charmer was originally designed and built by "Bernie" Pietenpol in the early '30s way up in the Minnesota north country.

Thousands of the "Piet", as it is affectionately called by builders and owners, have been built in all parts of the world. Many, many modifications have been tried through the years but rarely did any of them (airfoil shapes, etc.) result in improved flight characteristics over the original design.

This two seat high wing parasol is not fast but it is fun to fly and can lift itself out of most any small field.

The original design was intended for use with a converted Ford Model A engine. Many a builder had to build not only the aircraft structure, but the engine as well.

Over the years dozens of different engine types (auto, motorcycle, aircraft, diesel, etc.) have been installed in newly completed Pietenpols. The rarer engine installations included the 65 hp Siemens-Halske, LeBlonds, Anzanis, Velies, Kinners and Lamberts - to name a few strange names.

A large number of Piets have converted Corvair engines and, of course, the usual variety of small Continentals and Lycomings.

The **classical** installation calls for a genuine Model A Ford engine which might, just might, develop around 45 hp (wow!).

The airplane has a simple box fuselage made of spruce and plywood. The wing ribs are built up truss ribs with plywood gussets. The wing spars are solid spruce and the tail surfaces are simple, fabric covered wood frames.

This is one of the oldest homebuilt designs in the U. S. and it has a lot of admirers. Consequently, many close friendships have developed among the numerous builders and owners of Pietenpols.

In addition, there is an excellent exchange of information and parts, thanks to the efforts of the Buckeye Pietenpol Association (first formed in 1981) and its extensive membership.

Golda Cox

One of the all time "classics" . . . the Pietenpol Air Camper.

Perhaps the simplest of all aircraft to build . . . the Volksplane VP-1. Its functional lines have a beauty of their own. Note the large tires and simple single piece aluminum landing gear.

Their newsletter will delight the hearts of all builders, owners and historians interested in Pietenpol aircraft. Every issue is loaded with Pietenpol lore, building tips, material sources and latest flying activities.

THE VOLKSPLANE

The VP-1 single seater is, perhaps, the simplest and least expensive wood aircraft you can build. The plans first became available to builders in the mid-1960s. It was designed for use with any 1500cc to 2100cc VW engine. A two-place version, the VP-2 appeared in 1970. It required larger VW engines . . . anything from 1834cc to 2100cc.

The VP's, with their low fuel consumption of approximately 3 gph, are very economical to operate. Other than an occasional valve adjustment or an engine tune-up, the airplanes are virtually maintenance free.

The large 600 x 6 tires and a low stall speed of approximately 40 mph makes the Volksplanes ideally suited for grass strip operations.

Modifications are never encouraged by any aircraft designer but, as is the case with most designs, the VP-1 has also had its share. But, here again, attempts by the rank amateur to "improve" a good basic design seldom prove to be worthwhile.

If you want to build something simple and economical, the VP-1 is the plane for you. You can cut the construction costs even further by building up your own VW engine.

Although I don't consider the VP's to be cross-country airplanes, the one I built made the 1,200 mile trek from Texas to Oshkosh three times! No, I didn't do it - I couldn't sit that long on that hard plywood slab seat. It is, however, an ideal fun airplane for airport hopping and sightseeing.

HOW ABOUT SOMETHING MORE SOPHISTICATED?

In Europe they call them *touring machines.* These are generally two seaters equipped with enclosed cockpits, nice upholstery, and are mostly powered by small Continentals and Lycomings.

These all wood and fabric aircraft cruise about 125 mph or thereabouts. They can operate from unimproved airstrips because of their relatively low landing speeds . . . generally under 60 mph.

In this group of aircraft are a number of beautiful designs. Yes, they may be somewhat more complicated to build, but not always, as most of them utilize the basic box fuselage faired out with fairing strips. The thing that makes building one of these two seaters a bit more demanding is the fact that the wings may be tapered or have an eliptical shape with different sized wing ribs to build. Often the airplane will have a box spar for a one-piece wing . . . and that creates a problem for many builders who lack the necessary building space.

In this group of designs I would list Emeraudes, Minicabs, Jodel D-11's, Cavaliers and Jurca Siroccos. There are others, of course,

Most of these aircraft have been built in considerable number and the plans are still available. Check the classified ads in your SPORT AVIATION magazine to see who is selling the plans. Better still, ask anyone who has built one for details.

FOR THE FAST SET

Wood aircraft can be just as fast as any other kind. About all it takes is plenty of horsepower and money. The basic construction in many instances is no more complicated than that of the Pietenpol or Fly Baby. The complexity and cost stem from the tapered wings, retractable gear, controllable propeller and a cockpit full of avionics and electronics.

Here are three 200 mile per hour wood beauties you can build and/or lust

The naked beauty of the Super Emeraude design.

after: the Barracuda, Falco and GP-4.

Of this you can be sure . . . there will always be beautiful wood aircraft being built as long as there are homebuilders.

ADDITIONAL INFORMATION

FLY BABY - Plans $65, info $3. Peter M. Bowers, 10458 16th Ave., S., Seattle, WA 98168.

PIETENPOL - Plans information. Send stamped self-addressed envelope to Buckeye Pietenpol Association, 3 Shari Dr., St. Louis, MO 63122.

Also contact EAA - ask for the 1932 Flying and Glider Manual. It contains the original drawings for the two seater Pietenpol Air Camper. The 1933 Flying and Glider Manual has the plans for the single seat Pietenpol Sky Scout. Toll free number for ordering only, 1-800/322-2412.

VOLKSPLANE - VP-1 brochure showing what builders are doing - $4.50. Plans - $94. Evans Aircraft, Box 744-C, La Jolla, CA 92038.

Typical fin/rudder/empennage of an all wood airplane. Note the many reinforcing gussets.

METHOD OF SPLICING SOLID OR LAMINATED RECTANGULAR SPARS

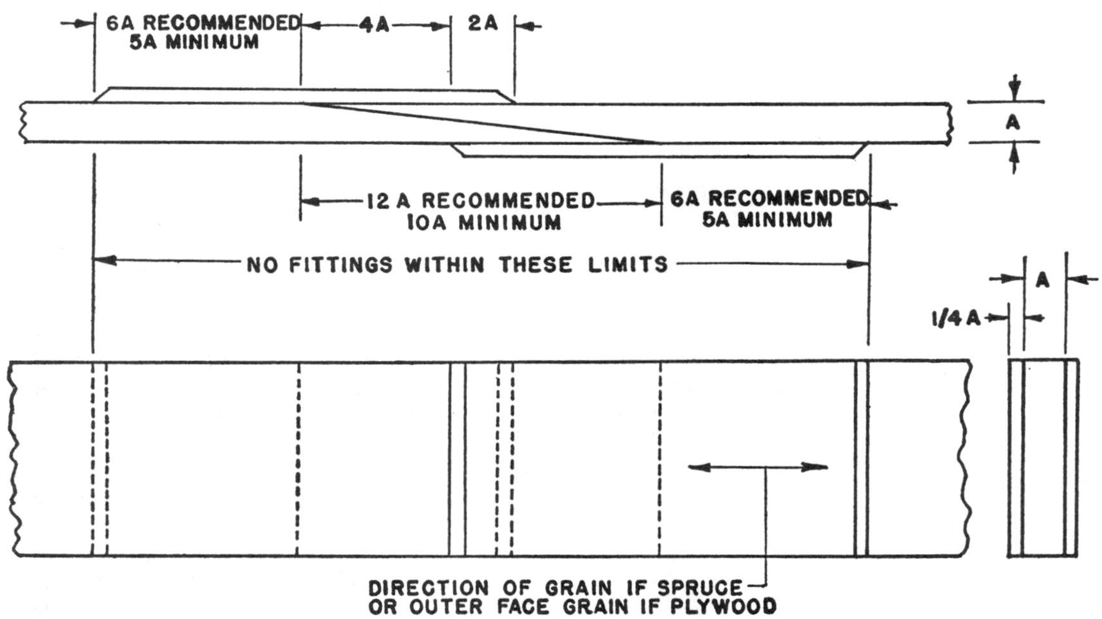

REINFORCEMENT PLATES TO BE SPRUCE OR PLYWOOD AND SHALL BE GLUED ONLY.

SOLID SPARS MAY BE REPLACED WITH LAMINATED ONES OR VICE VERSA, PROVIDED THE MATERIAL IS OF THE SAME HIGH QUALITY.

Why Aircraft Nails

By Arthur W. J. G. Ord-Hume, EAA 8579

WITH ALL DUE respects to the fine efforts of EAA home builders, there does seem to be rather a lot of emphasis on positioning brads or nails in wing-rib gussets, ply covering and so on.

I say undue emphasis because, as far as I am concerned, I ceased using nails for these tasks years ago. Brads are used by me where ply gussets or strips in excess of 1/16 in. (1.5 mm) have to be affixed. I also still use them on particular jobs where I am unable to apply glue setting pressure in any other way.

Before detailing my methods, it is as well to take a brief look at the reasons for using nails.

When casein and basic animal glues were used in aircraft construction, the glue line developed a rather indeterminate strength dependent on two points. Firstly, the joint had to be absolutely perfect so that a good surface-to-surface tight joint could be made "dry." The nail or brad was used to apply close-contact pressure between one thick member and one thin member, the best example being plywood to structural members. The second point was locating this close contact under predetermined setting pressure until the glue-line hardened.

The nail was thus required to provide and maintain setting pressure between two pieces of timber of largely differing thicknesses. It was also called upon to augment to a certain extent the strength of the joint both in glue-line shear and tension, although it is worth while noting that, in stressing a joint, the supplemental strength of nails was never allowed for.

With the advent of modern synthetic resin adhesives, considerable latitude is available in setting up and clamping a glued joint. Furthermore, the shear strength of a glued joint represents almost a constant between .002 in. and .070 in. thickness whereas casein demanded the thinnest glue line for maximum strength and any gaps automatically became "starved" due to evaporation and contraction of the glue substance on drying and hardening.

Today, we can make a perfectly satisfactory (but not recommended!) joint by laying one piece of wood gently upon another with a blob of glue between. We can glue a piece of ply to a spruce strip, provide the lightest of clamping for setting pressure, and still split the spruce in trying to peel the resultant joint.

From this, it is easy to see that the aircraft nail is as "old hat" as is the wire-spoked wheel.

The only features of the nail which we require today, then, are not those whereby it gives a clamping pressure and mechanical assistance to the glue line, but those by which it can **locate** pieces of wood to be glued until the glue is set. Once set, the nail serves only one purpose — to increase the weight of the structure. This last-mentioned it can do very well since on my first aeroplane I used nearly three pounds of aircraft nails which is just about equal to half a gallon of petrol or, if you like it another way, about 8 fpm rate of climb for that particular ship.

Accepting, then, that the nail is not necessary, what does one use? An obvious list of methods for applying setting pressure immediately springs to mind — the tacking strip, clamps, weights — or a patient hour or two with the thumb. However, the one to be used is the industrial stapling machine.

Tool for removing staples made from a screw driver. Heat the end, bend it through 30 deg., file a "V" notch in it, polish the underside and "heel" and re-temper.

Industrial stapling machines are available in a wide variety of shapes and sizes equipped to perform their duties in a variety of circumstances. The one I use is a conveniently-shaped squeeze gun embodying the following features:

a) Ability to use staples from ¼ to ½ in. leg length.

b) Two firing pressures regulated by thumb control.

c) Easily dismantled for cleaning.

With this tool, a gusset is fixed to a rib with three or four staples in a matter of seconds — remember that one staple is as effective as two brads or nails.

After a little practice, staples can be fired correctly at one a second and large ply panels on fuselage sides or wings can be fixed in a very short space of time.

The staples used are of flat-sectioned steel wire and are pre-pointed so as not to tear the wood fibres. Two types are available — ordinary hard steel and stainless steel. In cases where it will not be possible to remove the staples after setting, stainless steel ones should be used. Where they can be removed, ordinary ones are sufficient.

When the glue joint is set, the staples are readily removed with a shaped screwdriver (see illustration) thus leaving a clean wood surface with no nail heads to mar it.

Stapling saves time and gives a better job. Try it.

Construction Methods
WOOD Wing Ribs

by Tony Bingelis

BUILDERS OFTEN HAVE more questions about wing rib construction than some of the other more complex parts. It took 84 year old (yes, amigos, 84) EAA member H. Probyn of 93 Nyeri, Kenya E. Africa, to point this out to me. A review of my past mail log seems to confirm his observation.
NOTE: Readers might be interested to know that Probyn is an enthusiastic builder/pilot who has been flying since 1916. Currently he is 600 hours into a VW powered Jodel D-9 which he has built and flies about 100 hours annually. Having seen, used, and abused many wing ribs in his time, he believes that the readers would like to review wing rib construction with a look at jig requirements, costs, and the ease of construction of the different types of ribs. So, here it is.

The very first part most of us undertake to build is a wing rib. We can't all be wrong so it must be a good logical starting place for any project.

The skills required in wing rib construction are typical of those used throughout the remainder of the project. Ribs are economical to build and constructing them permits you to get the project underway with a minimum initial investment. Furthermore, building the ribs enables you to gain the feel and experience needed for other construction. If you later find you don't like to do that sort of thing, or cannot discipline yourself to a rather long period of turning out wing ribs, it could be the project is not for you. That, maybe, you shouldn't be building the airplane in the first place. With no large expenditure of funds to influence your decision, you can gracefully bow out at this point.

It is pretty hard to build a wing without utilizing some sort of ribbed foundation structure. The nearest thing to a ribless wing would be one of composite construction. But, even here, airfoil sections akin to the lowly rib must be cut out of some sort of foam material usually, to an exact dimension, shaped, and then assembled prior to skinning the wing.

When you delve into construction methods you find that just about all imaginable materials and techniques have been tried, at one time or another, in wing rib construction. Aluminum ribs hammered out of a single sheet, built-up aluminum ribs riveted together from preformed aluminum strips and extrusions, steel rod ribs with silver soldered or brazed joints, built-up wood ribs and solid wood ribs, and of course a variety of plywood ribs. As you might expect, there are countless variations in design for each of the types of construction mentioned.

JIGS FOR WOOD RIBS

Ribs require some sort of a jig during their construction to insure uniform contours for all ribs.

Whenever a tapered wing design of any kind is selected, the builder should be aware that his jig requirement and rib building effort will cost him extra time and money. As a starter, it would be necessary to lay out and build a separate jig for each size rib. Lucky are the ones who are building a rectangular wing design as they will only need to make one rib jig.

If you only need to build but 2 ribs in each jig, it is unnecessary to build as fancy and as durable a jig as you would if 20 to 30 ribs had to come out of that same form.

After the rib shape is accurately drawn on a piece of ¾" plywood, overlay it with wax paper or a sheet of polyethelyene to prevent glue from sticking things you don't want stuck. Use headless nails to outline the rib shape and to locate the positions for the uprights and diagonal braces. Small blocks of wood can also be used.

Shown are a solid wood rib, a reinforced plywood web rib and a typical truss rib.

Several ribs in the process of having gussets glued in place. Note that many clamps are required.

These blocks should be a bit smaller to dimension than the rib stock used. You really don't need too many nails or blocks. Just locate them whenever necessary to hold the rib stock to the exact curvature required.

If you do not intend to glue-nail the gussets (assuming your rib type requires gussets) but prefer to save weight by gluing and clamping them only, a slight modification in the jig board and procedure is suggested (See photo No. 1). First, cut the jig board to the approximate shape of the rib reducing its overall size as much as possible. Then, nail several small blocks (about ¾" x 1½" x 6") across the bottom of the jig board so that it is elevated above your work tables. This setup permits you to place clamps on each gusset as it is glued in place. The drawback to the glue and clamp method, of course, is that it will be necessary to leave the rib in the jig for at least 6 hours, preferably overnight. And then, the same delay for the attachment of the opposite side gussets.

The new miracle glues that dry almost instantly might change all this. However, before using them you'd better check them out for water resistance and durability under adverse heat and cold conditions first . . . then check on your bank account.

Another protective treatment for your jig board consists of drilling a large hole 1" to 1½" in diameter at each joint intersection of your rib jig. The idea here is that if there is no wood there, the rib cannot stick to the jig and removal of the finished rib later will be a snap (oops). No need to use protective wax paper on such a jig.

To be assured that your spar openings are properly sized and shaped, secure a block of wood of the same dimension as the spars to the jig board.

The severe curvature at the nose end of most ribs dictates that the capstrips first be preformed by soaking or by steam bending. Soaking the capstrips in hot water for about 30 minutes and then clamping them in a bending fixture similar to that illustrated in Photo No. 2 is usually sufficient for all but the curviest of ribs. To allow for spring back, cut the bending jig (from scrap 4" x 4" block) to a curvature slightly greater than the rib shape at the nose end.

Before loading the capstrips in the bending jig, flex each one to see which way it bends easiest. Be sure to allow the bent capstrips to dry thoroughly before removing them from the fixture and doing any gluing. 24 hours should be sufficient unless the humidity is high.

QUALITY CONTROL

After the ribs are removed from the jig, there is really no need to get them all strung on a short stub spar and sanded with a sanding block until all are exactly alike. By the time they are glued to the spars, you, yourself, may well have introduced slight misalignment in a rib or two. The wing should be sanded after assembly, with a long 4' sanding block to level everything . . . so, why do it twice?

The ribs should be trimmed and the spar openings cleaned up of glue runs . . . so should the capstrips at the gusset areas. Your best tool for accomplishing this is not sandpaper but a 10" bastard cut file. If you still prefer using sandpaper after trying the file, be sure to use a sanding block so that you don't round the areas you don't want rounded.

A file is a handy wood working tool. For example, after you have slipped the capstrips and the uprights and diagonals into the jig for gluing, you can file lightly over each joint to be sure each piece is level with the other. Always level your joints before applying skin or gussets. Always.

TRUSS TYPE RIBS

These are the ribs built up of strips of spruce, white pine or some other type of acceptable wood stock. Truss ribs are usually reinforced at their joints with small plywood gussets. These ribs when judged on a strength to weight basis are always the lightest of the rib designs, if properly designed and well built. Unfortunately, they are also the most time consuming to build.

Early in the 1930's a gent named Les Long used a truss rib featuring dual capstrips in an ultra-light design called the "Longster". At first, you might think such a rib would be very heavy but this was not so as that truss rib did not have plywood gussets. It was an easy rib to build although it did require more spruce stock. Using ¼" x ¼" spruce strips, he slipped two of them into the jig for the top part and two for the bottom. Separating the two strips slightly he slipped in the spruce uprights and diagonals between the cap strips. A dab of glue where needed and a single ¾ x 20 gage nail driven through each joint completed the job. That was it. The rough ends sticking beyond the capstrips were then trimmed with a model saw and the rib was finished. Supposedly, one of these ribs could be turned out in 10 minutes. It weighed a mere 3½ ounces (54" chord). Maybe builders should take another look at this design. You certainly can't build a regular truss rib in 10 minutes.

The order of efficiency (weight compared to strength) for the different rib types might be rated as follows:

1. Truss rib
2. Lightened, reinforced plywood rib
3. Solid plywood web with stiffeners
4. Plywood web with lightening holes and no reinforcements
5. Full plywood web

Capstrip bending jig is used to preform as many as 4 capstrips at one time. It is cut from a single piece of 4 x 4.

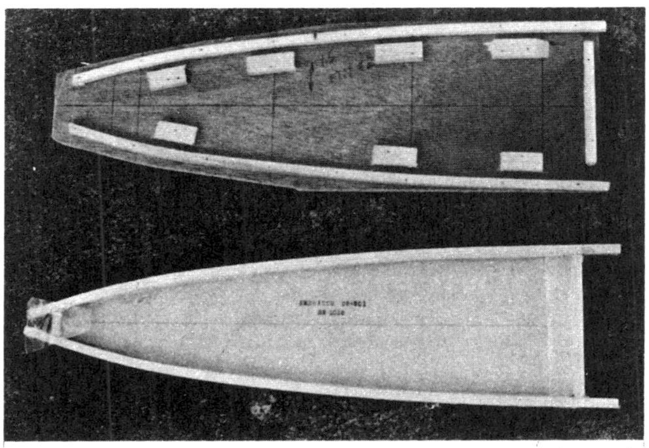

A plywood rib (lower) has just been removed from the forming jig shown above with its blocking. Note scrap of wax paper still stuck on the rib.

While well designed ribs will generally reflect the foregoing order of strength to weight standing, the truss type may not always be the best type of rib to use. For example, when shallow or rather thin ribs are required, it is better to use the plywood web type. This applies as well to some slim wing tip ribs which do not lend themselves to the truss type of construction.

Truss ribs aren't too effective when used under plywood skins. The gluing operations are ordinarily accomplished by means of nail strips. When the nails are hammered through the skin, the capstrips being quite flexible will give and deflect in the process making that hoped for good tightly clamped skin joint somewhat questionable. Truss ribs, however, built with long one piece plywood gussets could certainly minimize the flexibility of the capstrips and make the rib as suitable as a solid web type.

If you glue-nail your gussets and you are using 1/16" plywood gussets, ¼" x 20 gage nails driven about ½" apart will suffice. Use at least one nail in each upright or diagonal. Half round or quarter round gusset shapes prove the most effective and can be easily cut out in number with a 2" hole saw. Stack a bunch of them at one time for cutting into halves and into quarter shapes. The drilled hole left by the hole saw can be trimmed away or ignored. Always lightly sand or scuff birch plywood before gluing it.

PLYWOOD RIBS

In this category, we include the full plywood web ribs, with or without stringers, and the plywood web ribs with lightening holes. I guess it is pretty much true that the heavier the rib the stronger it is. However, just how strong does a rib have to be? Rib failures are really a rather rare occurrence.

Plywood web ribs, although considered to be among the heavier of the rib types are easier to construct, hence quite popular with the beginner builder.

Whenever lightening holes are to be spotted on the rib layout, remember that one large hole will lighten a rib more than two smaller diameter holes. See Fig. 1.

The hole arrangement and size are critical since most plywood rib failures usually occur around these holes. Whenever in doubt as to the strength in a particular area, use stiffeners close to the cut out.

Locate lightening holes so that they run longitudinally. Generally speaking, ribs with the face grains of the plywood webs oriented vertically, consistently demonstrate greater strength when a full solid web is used. Longitudinal face grain, on the other hand, is better when webs are lightened with holes and when stiffeners are used. Small stiffeners glued near the edges of lightening holes are very effective in reducing buckling. Fortunately, the resulting increase in weight will often be offset by the increase in strength . . . but only if that added strength is really needed.

In some antique aircraft, you will find plywood ribs featuring a single capstrip made of ¾" x ¼" spruce with a shallow groove to receive the plywood webs. Another common treatment was to glue and nail capstrips on each side of a plywood web. Both of these construction methods are still used to some extent but the current trend is to use a ¼" plywood web rib without any reinforcement around its periphery.

Years ago, most builders had to cut out plywood ribs with a small coping saw or jig saw . . . a very slow way indeed. Today, a band saw permits a more accurate cutting of the ribs and speeds the process remarkably. The lightening holes are cut out with sweep cutters and with large hole saws that can be chucked in any drill press. (Oh, my gosh, no! . . . never in a hand held electric drill). A lot of builders are also using routers to cut out both the rib and the lightening holes too. An all-in-one jigged operation is possible with a router.

THREE PIECE WING RIBS

Some ribs are built in three sections with a separate piece for the nose, the center section (between the front and rear spar) and the trailing edge portion. In assembly, each of these pieces is butted against the appropriate spar and nail-glued in place. This type of construction is used to permit building the spar to the maximum wing depth for strength. The alignment in assembly of this type of rib is critical as any deviation in the vertical uprights at any one of the rib ends could alter the incidence of the wing and/or cause assembly and alignment problems affecting the external contours of the airfoil.

It is suggested, therefore, when the plans call for three piece wing ribs, that they be built in one piece in the jig and then later cut apart at the spar openings before assembly.

Each upright that butts against the spar must exactly match all of the other ribs to insure alignment.

HOW DO YOU ATTACH THE RIBS TO THE SPARS?

One piece wing ribs are simply slipped on to the spars to previously marked locations and nailed and glued into place. The rib openings for the spars must not be a press fit. Allow .010" to .015" tolerance for glue and the swelling of the wood from the moist glue.

Do not drive nails through the top or bottom capstrip when attaching the wing ribs to the spars. The proper attachment is always by means of glue, using cement coated or brass aircraft nails driven through the rib upright member, on each side of the spar. A couple of nails through each upright should do the job.

When corner blocks are used, try gluing and nailing one of the corner blocks to each rib location on the spar before trying to attach any of the ribs. Positioning the rib after that becomes an easy job as you'll have a solid fixed backup against which you can easily position the rib and fasten it in place without it sliding all over the place while trying to nail it.

When gluing the individual rib sections of a three piece rib to the spars, hold a block of wood against the spar and rest the bottom of the rib section against it for alignment. See Figure 1. Or, you might try this for aligning any kind of rib. Mount the inboard and outboard wing ribs first. Then span across them with a long straight-edge set on top of the ribs near the front end. Place another straight-edge near the aft end of the ribs. As you position each rib for gluing make sure that it touches both straight-edges. Alignment is assured. Be sure all ribs are slipped on in their proper locations before permanently attaching the end ribs!

Assure yourself that the front straight-edge and the rear one are parallel to the main spar when viewed from the front or rear of the wing unless you desire to build in some degrees of wash-in or wing warp.

HOW ABOUT COSTS?

Well, we know that, of the wooden ribs, the truss ribs are lighter and more efficient and that the plywood slab rib is easiest to make and that the plywood reinforced rib is somewhat heavier, but also usually stronger . . . but how do they rate cost wise?

Figuring the cost on a comparative basis for a typical 48" wing rib, we find the rib costs work out to approximately:

Truss rib — $1.53 — (including spruce/plywood gussets/glue/nails)

Les Long spruce rib — $1.18 — (including spruce/glue/nails)

Solid ¼" plywood rib — $2.34 — (aircraft mahogany plywood)

Reinforced plywood web — $2.68 — (⅛" aircraft plywood/spruce/glue/nails)

Reinforced plywood web (both sides) — $3.52 — (⅛" aircraft plywood/spruce/glue/nails)

NOTE: You might do better or worse price-wise depending on your plywood and wood costs.

Apparently, there is no single best rib design for all purposes. Rest assured however, that your designer did consider all the advantages and disadvantages of the rib type he finally selected. If you are designing your own original, perhaps this discussion has given you a firmer grasp of the subject.

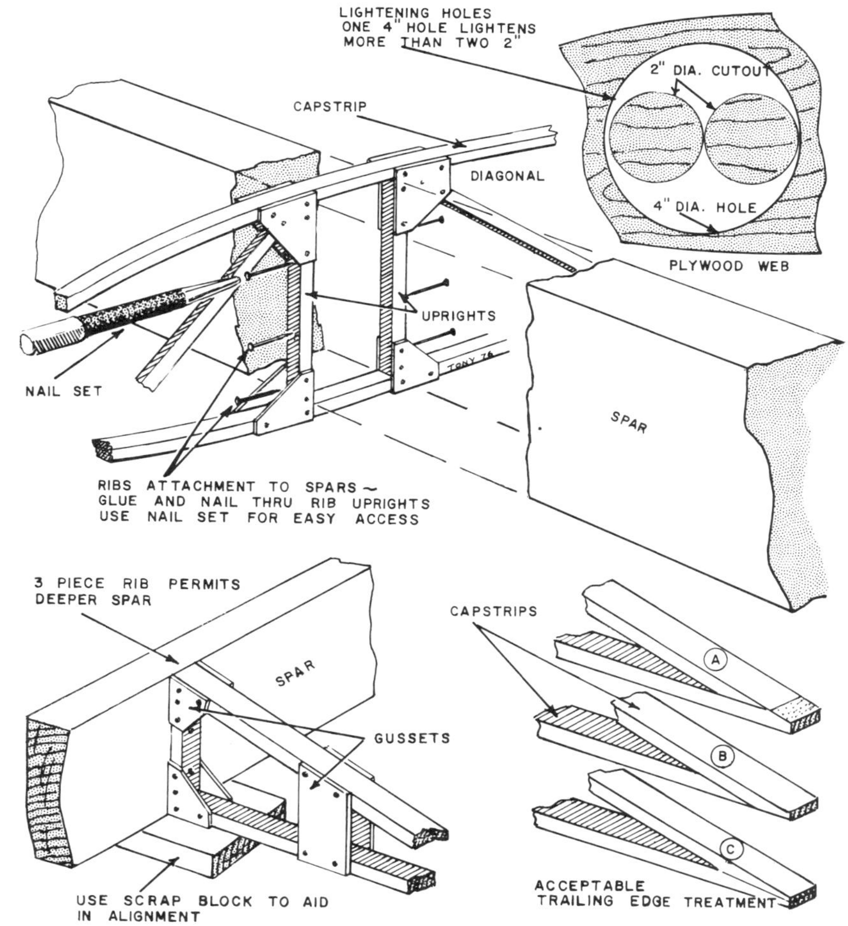

FIGURE I.

Typical truss ribs with mahogany gussets glued and nailed. Builder was a bit generous with the nails. Looks easy enough to build doesn't it.

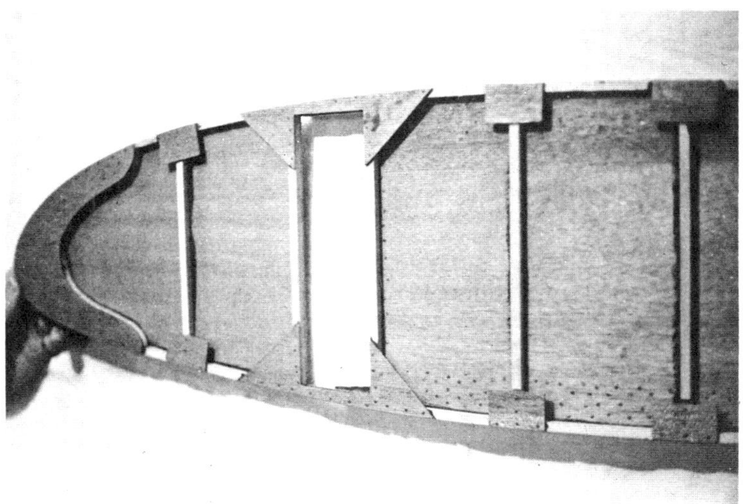

This plywood web rib is reinforced and has gussets on the back side. Rectangular gussets could have their corners cut off as the overhang serves no purpose.

Aircraft Plywood . . . How to Use It

by Tony Bingelis

WOOD HAS BEEN an important material in aircraft construction since the days of the Wright brothers, and those of us who build and fly aircraft with wood components remain convinced that in spite of some very real disadvantages, wood is an excellent construction material, well adapted to the technical expression of modern aircraft design.

If you think about it, you can easily see that the construction of wood aircraft would become very complex, if not impractical, without the use of plywood. As with any other construction material wood, and particularly plywood, has peculiarities and characteristics which must be taken into account during and after construction. The more you know about it the better your craftsmanship should be.

PLYWOOD IS PLYWOOD BUT . . .

Take two or more thin sheets of wood veneer, glue them together with the adjacent grains running in opposite directions and you will have made a sheet of plywood. As the strength of wood along the grain averages about 18 times as great as its strength across the grain, it can be readily understood that the more cross plies of veneer used, the more constant will be the strength ratio in all directions of the finished product. The strength of the plywood depends upon the type of wood used for the plys and on the characteristics of the glue bond.

As I see it, there are three types of plywood. The lumberyard variety, the marine grade plywoods, and the aircraft-quality product. As you may suspect, the lumberyard plywood is considerably cheaper than aircraft plywood or marine grade plywood. Aircraft and marine plywoods are made utilizing waterproof glues and, although the common exterior grade plywood is likewise made with waterproof adhesives, there is a very important difference in each variety. The difference is in the degree of quality control exercised in manufacture and, of course, strength.

Commercial grades of plywood are glued together utilizing veneer scraps and pieces for the inner core. Very little attempt is made to fit these pieces closely and, as a result, almost every sheet you buy will have internal gaps and voids, some of which are quite visible from the outside edges of the sheet. These voids seriously reduce the strength of the plywood sheet and, its use for structural components or stressed skin applications is not recommended.

A somewhat greater control of quality is exercised during the fabrication of marine grade plywood but, it too, is very likely to contain occasional small voids.

Aircraft quality plywood is manufactured under the strictest standards in the industry and no voids are permitted. Another bit of difference, of course, is that aircraft plywood is commonly obtainable in 4' x 8' sheets as thin as 1/32". The 1/16", 3/32" and the 1/8" plywood sheets constitute the most popular sizes used in the average aircraft project.

BEAUTY IS ONLY SKIN DEEP BUT . . .

Some plywood covered fuselages and wings do look terrible. The builders may have done everything exactly right and by the numbers but . . . if the big gluing jobs happened to have been done in the winter time in a heated workshop, chances are that the plywood skins will have acquired an ugly appearance by spring time when the structure was subjected to increased humidities.

The normally dry winter air is made even more so when a space heater or a furnace is used to maintain gluing temperatures and comfortable working conditions in the workshop. If the house and shop have no humidifier, the artificially heated air will become exceptionally dry and the plywood will be subjected to abnormal shrinkage. Then, summer comes with its warm humid days. The plywood swells with the increase in humidity but, since the plywood has been glued to the aircraft's structural framework, the plywood edges are permanently secured. Something has to give as the temperatures and humidity increase. And, what gives, is the plywood skin by bulging inwardly or outwardly. Since all of the structural bays are not of the same size and shape, the stress developing is not uniform throughout the plywood skin and it takes on the appearance of undulating waves . . . ripples and bulges that are all too noticeable. This effect is certain to occur when thin plywood skins are glued on structural components that have little or no curvature. Flat fuselage sides, for example.

Fortunately, the problem is more cosmetic than structural although it would seem that a slight reduction in strength would result in highly stressed areas where severe bulges are present.

Unfortunately, there is nothing that can be done after the bulges appear. However, an abnormal swelling or shrinking of the plywood can be minimized if:

1. You do the gluing when temperatures and humidities are closest to the summertime norm.
2. You reduce the size of unsupported panels. Obviously the larger the unsupported area spanned by a thin plywood web, the greater the sag potential.
3. You lay plywood diagonally even though it will take more plywood to do this. Naturally, 45° plywood can be used but it is more expensive than the regular 90° grain sheets. Either way it would cost you more.

USE IT JUDICIOUSLY

Plywood has three distinct applications in aircraft construction.

1. It is used to strengthen and unify structural joints (gussets).
2. Plywood sheets are bonded in place to serve as external stressed skins on wing surfaces and fuselages.
3. It is used in stressed areas where bolted metal fittings are installed, to reinforce the area and to diffuse concentrated loads.

Plywood prices are no longer attractive as they once were and even the small single seat all-wood aircraft will require several hundred dollars worth of plywood. With costs like that, efficient use of plywood becomes very important to you the builder.

In the interest of economy, therefore, cut out all of the large skins and pieces from the plywood sheets first. Resist the temptation of cutting off a piece or a corner so you can get started on some small parts such as ribs and gussets. Furthermore, always allow for trimming afterwards . . . don't attempt to be foolishly frugal as you may find yourself wasting more plywood by making parts over again.

Mahogany and birch plywoods are the two most commonly used materials by homebuilders. Mahogany is the easiest to glue and is the lightest of the two. However, birch is stronger than mahogany and may always be substituted for mahogany with no increase in size necessary. Do not substitute mahogany for birch plywood in the same thicknessess unless you're sure you know the decrease in strength is not a factor.

Plywood is heavier than spruce and must be used judiciously in order to keep the weight of the structure down. After all, a joint will not necessarily be strengthened or improved by making the gussets larger. Chances are that even a small gusset can do an adequate job. Take a look at Figure 1. to see how this works.

SIZE OF UNSUPPORTED PANELS

In theory the maximum size for an unsupported plywood panel is dictated by the amount of computable stresses. Plywood panels which are subjected to the highest shear loads are those utilized in spar webs, D-section nose skins, and fuselage side panels. In most other areas the design loads are light or insignificant. A designer will usually select a combination of panel sizes and skin thicknesses to meet anticipated handling loads, appearance and aerodynamic smoothness requirements. In some areas thinner skins may safely carry any load which may be imposed but he realizes it may be more practical to use a slightly thicker sheet for appearances sake. The amateur builder is not encouraged to arbitrarily increase plywood skin thicknesses. For example, indiscriminately substituting heavier skins in the outerwing sections may, in effect, stiffen the outer panels and tend to transmit greater bending loads to the center section of the wing, which may not be able to handle the added concentration of stress. Remember, you're most likely to achieve safe results if you stick to the proven design . . . particularly in the important structural and load carrying components.

GRAIN ORIENTATION

The direction of grain of the outer surfaces of the plywood is very important in some locations and relatively unimportant elsewhere. If your plans indicate a specific direction for the grain, by all means comply. If no grain orientation is shown on the drawing, you may assume that the option is yours.

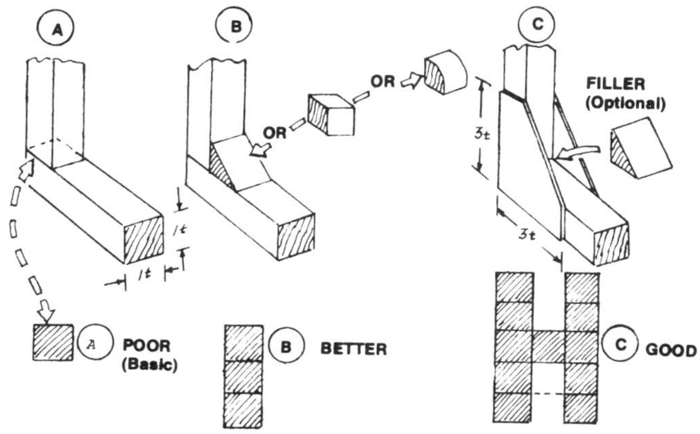

FIGURE 1.

Within reason . . . THE GREATER THE GLUE AREA, THE STRONGER THE JOINT

PLYWOOD PREPARATION

The first caution that comes to mind is that plywood, particularly birch plywood surfaces, must be lightly sanded to remove the glaze in the areas to be glued. Mahogany is easier to glue as it has a comparatively coarse porous surface. Birch, on the other hand, is a dense fine-grained hardwood that becomes even more difficult to glue as a result of the pressing process involved in the plywood's manufacture. A hard surface sheen is imparted which is very resistant to the penetration of most glues. This inability of glue to infuse itself into the glazed wood fibers can result in inadequate and unsafe joints. Therefore, the slick surfaces of plywood should be "broken" by a light scuffing with sandpaper, or by scraping them gently with a sharp wood chisel or plane blade. Test for yourself the difference this preparation makes. Spill a little water on an unprepared scrap piece of plywood and also some on another lightly scuffed piece. Notice how the water stays in droplets on the unprepared surface while the other piece becomes thoroughly wetted.

Never neglect the preparation of plywood for gluing. When doing this you can save time by sanding large pieces of plywood stock before cutting it into gussets and other small parts rather than sanding each little gusset separately before gluing it in place.

When preparing large plywood skins such as fuselage sides or wing skins for gluing, it is a common practice to drive a headless nail in each end of the plywood to hold the plywood in alignment and to keep it from moving while you trace the outlines of the internal structure on the skin. You will find it most helpful to also draw the outlines of the internal structure on the outside of the skin. This will prove to be most useful during the construction of the airplane for the subsequent attachment of parts, fairings and the like. I would strongly suggest you **not** use a ball point pen with indeliable ink as the stuff will often bleed thru surface varnish, paint and fabric. It is far better to use a pencil for the marking process.

PREPARATION OF THE STRUCTURE

Prior to skinning, use a straight edge (a two inch wide board three or four feet long will do) to aid in truing the structure. Lay the board across the surface or, in the case of a wing, across at least 3 ribs, and check for low spots. An easier way to do this is to rub colored chalk along one edge of the straight edge. Then, when the straight edge is rubbed across the structure, all of the high spots will be smudged from the chalked straight edge rubbing on them. A gentle sanding of the high spots with a sanding board will soon level the structure to perfection. When all of the structure displays a uniform smudging from the chalk rubbing . . . you have it.

Finally, before gluing the plywood skins to the struture double check the structure for double curvature. Compound curves cannot be covered by a single sheet of plywood (or aluminum for that matter). If you suspect that there may be some compound curvature in the structure, test the surfaces by using a sheet of stiff paper or light cardboard, formica or similar material. If its outer edges lay in place evenly without bulges and ripples forming . . . the surface is O.K. No compound curvatures are present.

GLUING AND CLAMPING

Most builders have switched to the epoxy adhesives the majority of which are effective, easy to use, and do not require the heavy clamping pressures that the Recorcinol glues were noted for. Plastic resin still prevails in popularity as an economical and popular glue but, it too, requires a moderate amount of clamping pressure. While clamping pressures are easy enough to achieve around the perimeter of a large sheet, you must resort to staples, nails, nailing strips and weights for the application of uniform pressure across an entire sheet of plywood. There is no great advantage to permanently nailing plywood in place. Nails and staples should always be removed from the external surfaces of the aircraft as they have a nasty characteristic of protruding and ruining the effect of a fine finish on the aircraft. Elsewhere, leave the nails if you prefer.

The previous placement of a couple of locating nails will now assure you exact realignment of the skin when gluing it to the structure.

Be sure that the glue areas are not contaminated by varnish. Scrape off any offending areas using a sharp wood chisel or block plane blade held vertically. Also be sure that the glue areas have been sanded, dusted, or blown clean and are ready for gluing.

BENDING PLYWOOD

Plywood can be bent dry or it can be preformed by the use of heat and moisture. Where the bend is shallow, as over the top surface of a wing, the plywood is merely bent in its dry state and glued in place on the structure. Where the bend or curvature is severe, as around the leading edges of the wing, flaps, or ailerons, the plywood should first be soaked in hot water, at least 20 minutes, and then formed around a hot metal pipe of the correct diameter. To do this, a metal pipe may be placed across two saw horses and heated with a welding torch, or by whatever other means is available to you. Prior to soaking the plywood piece in hot water, (a tub or shower is a good place if someone isn't in it at the time), draw a pencil line along the centerline of the proposed bend. That pencil line will be used later as an alignment reference during the bending process. If the plywood piece is large, (3' to 4' long) it might be easier to bend if you had somebody else helping. It can be done alone, though.

Place the wet hot soaked plywood over the pipe. Align the reference line with the centerline of the pipe and slowly and smoothly force the plywood down with your hands, around both sides of the pipe. In most instances, you will find it best to bend the plywood well beyond the angle required as when you release the pressure the plywood will have some spring-back. It will also tend to straighten, somewhat, during drying. To insure

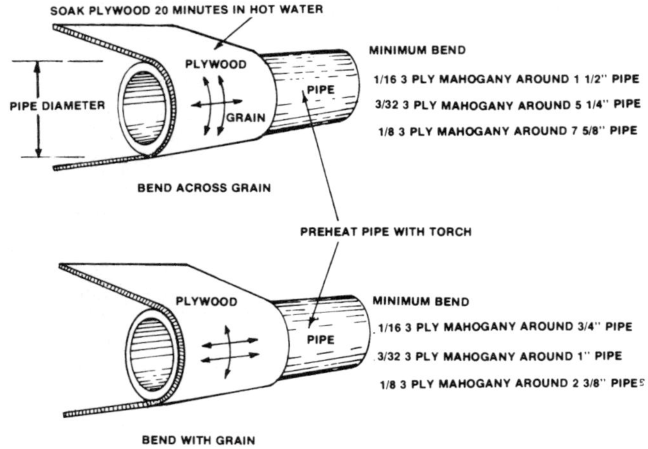

FIGURE 2.

that the plywood will have sufficient bend after drying, tape the ends together or slip a spacer board between them and tape the plywood to it while the wood is drying. Allow the preformed plywood to dry at least 24 hours before gluing it to the structure.

If you have never bent plywood around a hot pipe before, you will be amazed at the ease with which you can achieve a nice even permanent bend. The hot pipe literally steams the wet plywood into shape. It may be necessary to re-wet the plywood with a sponge during the bending operation if the pipe is extremely hot. It can char the plywood unless it is moistened a bit.

As an approximate guide to the minimum radius to which three-ply plywood (mahogany) may be safely bent, use the information shown in Figure 2. It is well to accept the fact that bending plywood in its dry state always requires a much larger bend radius than if it were first wet soaked . . . as much as 8 times larger, in fact.

PROTECTION AGAINST DETERIORATION

Well ventilated dry wood will last for centuries. Proof of this is in Old Ironsides, a relic of American History. It still floats sedately in its Charles river harbor near Boston. Of course, wood will deteriorate, but so will most metals. Naturally, there is a difference in the means taken to prevent deterioration. In both instances, the biggest potential problem is from entrapment of moisture which, in time, may result in wood rot, or in metal, corrosion. Rot, like corrosion in metal, can eat away the material's strength until it ultimately fails under the slightest stress.

The best protection you can provide is to insure a well ventilated structure with an adequate number of drain holes located in the lowest areas. In addition, there are good wood preservatives such as copper napthenate and pentachlorophenol which will prevent or retard deterioration of properly impregnated wood. This treatment followed by two or three coats of a good polyurethane varnish applied to the structure helps seal the wood and reduce its sensitivity to humidity changes.

Another good surface treatment and weatherproofing is to "varnish" the plywood and the underlying structure with T-88 Epoxy glue thinned with lacquer thinner to a brushing consistency. This is also used to weatherproof the entire structure, a method used by some of the country's finest wood boat builders . . . not flying boats, amigo, just boats.

. . . from Tony's Notebook/PLYWOOD

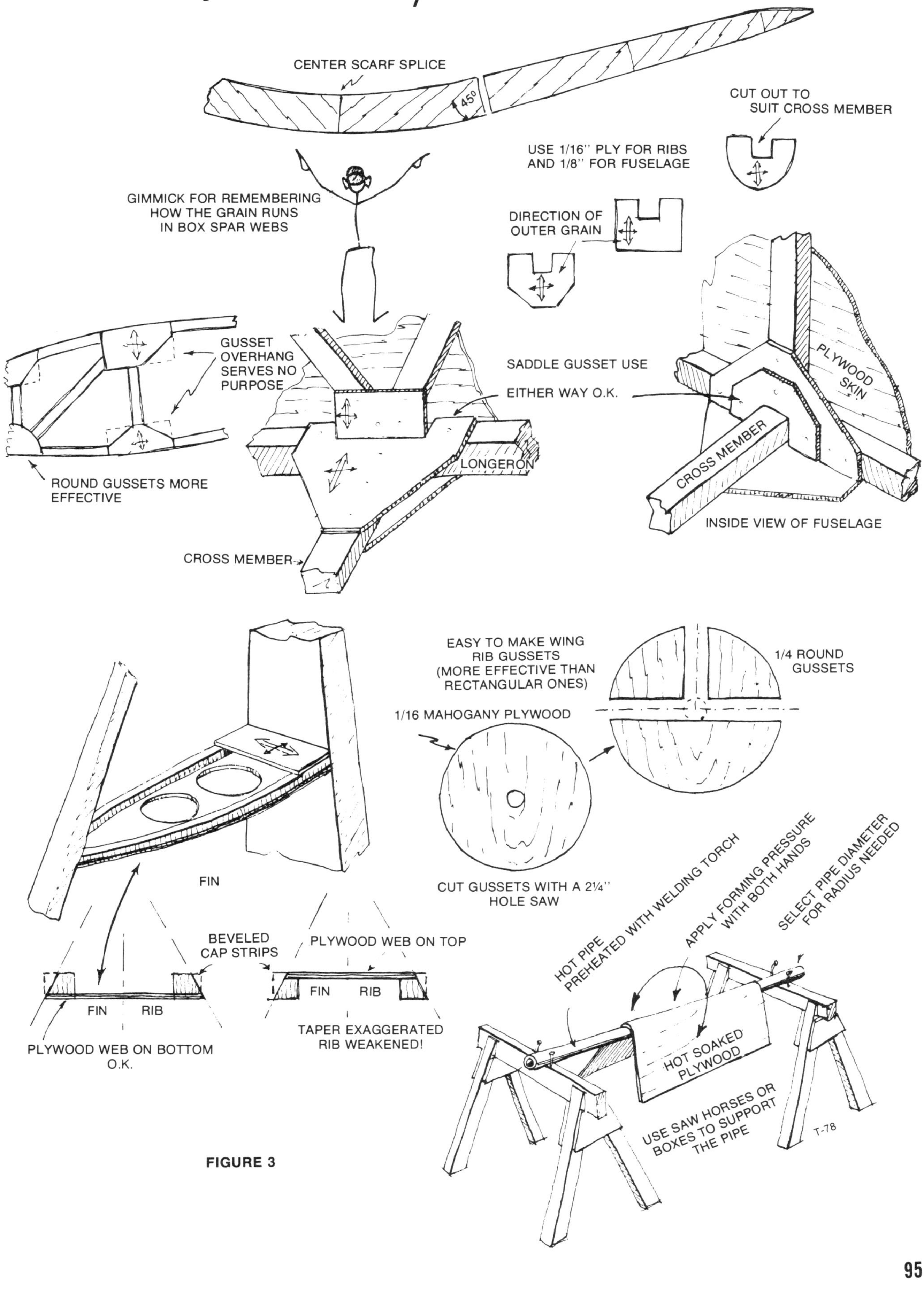

FIGURE 3

To Nail ... or Not to Nail

by Antoni (Tony) Bingelis

IF YOU ARE building in wood you probably have already decided whether to use nails or not. Undoubtedly, your decision was, to a great degree, influenced by personal preference as well as by the design details. If that is the case, certainly, there is no intention here to influence you to change your thinking on the subject. Be assured then that this impartial (chuckle) commentary may be just the thing for somebody else.

NAIL SELECTION

In determining the nail type to use, I would give preference to brass nails over ferrous types every time. This is true even though steel nails, as manufactured for aircraft use, will have been specially treated by cadmium plating, galvanizing, or cement coating to prevent rusting. Equally important is the selection of the proper length and size of the nails to use.

Nail sizes are expressed in wire gage sizes for the diameter and in inches for the length. In most ordinary sizes the more slender 20 gage nails are recommended for use as the thicker 18 gage size is heavier and more apt to cause the wood to split.

An aircraft nail to hold effectively must be long enough to penetrate into the underlying structure to a minimum depth roughly equivalent to 3 times the thickness of the surface plywood. For example, in gluing and nailing 1/16" thick gussets to wing rib cap strips, ¼" x 20 gage nails would suffice. These ¼" nails would pass thru the gusset and penetrate the structure 3/16". Nails shouldn't be so long as to pass completely thru an underlying structure.

NAIL HANDLING

Since most nails used in aircraft woodwork are small and difficult to handle with the fingers, a number of handling gimmicks have evolved. Each of these methods is lauded by some devoted adherent as being the best way to pick up nails and to hold them in position for nailing.

Some advocate the use of a magnetic tack hammer even though this method requires a steady hand and a good aim in order to land the nail exactly where you want it. It may be alright to use this technique when a random nail or two is to be driven to keep a panel from sliding around as you prepare to secure it, but rapid and accurate nail placement will be difficult for the average builder. (Naturally, a magnetic hammer will not pick up the brass nails.)

Another source seriously proposes that the plywood be pricked with a sharp awl or a pair of dividers and the nail then inserted, by hand, in the hole just generated . . . obviously a slow and crude procedure.

Still another way described in some ancient reference goes something like this. Cut a long notch in the end of a slender stick. Pick up a nail and slip the nail's shank into the notch with the head up. The stick can then be moved into position and the nail, without any risk of damage to the fingers may be bravely hammered home.

Well, the method which merits my personal favor has several variations. First, scatter some nails on your work surface within a 3 or 4 inch area, and then using a pair of tweezers, (small nosed pliers, or even a small magnetized screw driver blade would be O.K. for steel nails) pick up *a nail*, hold it in position, give it a light tap with a tack hammer, and you are on your way. Incidentally, most builders can handle the ½" or larger nails easily enough with their fingers.

Your own practice run on a couple of the methods described, or a variation thereof, will help you determine the most effective method for you.

PERMANENT NAILING

Permanent nailing as the term implies is a method of applying clamping pressure on a glue line thru the use of nails which are not later withdrawn after the glue has dried. Instead the nails become an integral part of the structure. Permanent nailing is used by many builders as a quick and convenient way to affix the various plywood gussets to wing ribs,

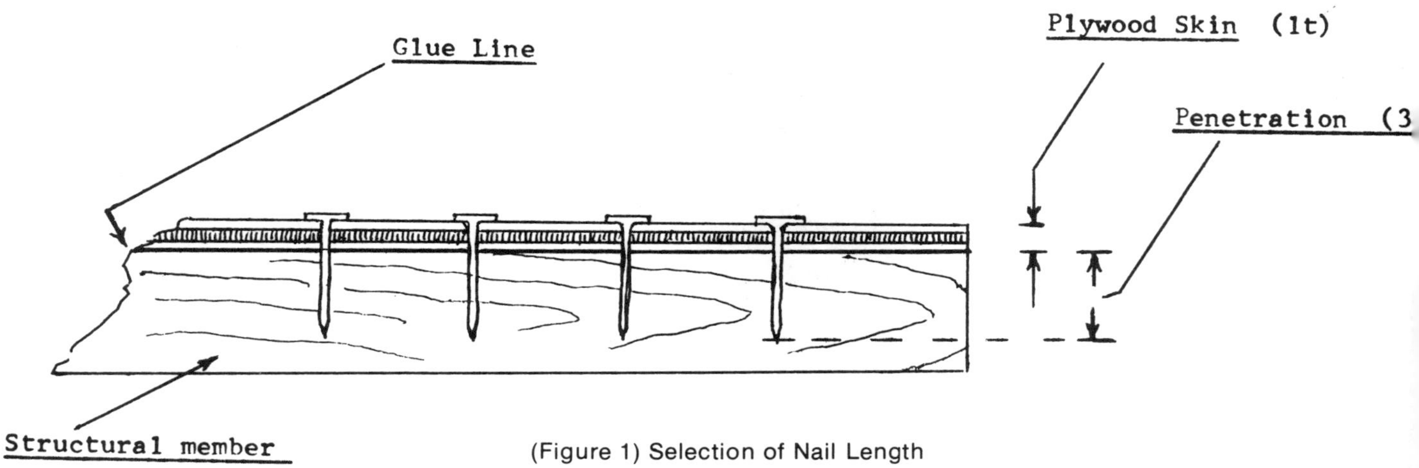

(Figure 1) Selection of Nail Length

fuselage frames, and control surfaces. Quite often, nails are even employed in the application of plywood skins to major structural units.

Any place nails are permanently driven in external surfaces, or wherever appearance counts, use strips of ducting tape or pinked tape doped down over the nail rows before the final covering and finish is applied. However! *Before* using permanent nailing to any extent in any external plywood skinned surfaces, think it over thoroughly, and then . . . decide against it. Elsewhere permanent nailing can be useful and quite effective.

Even in projects where the intention is to clamp all gussets and permanent nailing is not desired, it may still be found expedient to drive a nail or two into the gusset to hold it in position while you are setting up the work for clamping. The nails will guarantee freedom from worry that the gusset will slip out of position as the clamps are tightened.

The larger plywood pieces and gussets require pre-nailing in order to make the nailing go quickly. However, pre-nailing of plywood skins and gussets in thicknesses less than 3/32" is often more trouble than it is worth as some of the preset nails will drop out because of the vibrational shocks created as the nail hammering process continues.

Because the pressure exerted by nails is rather low, uneven, and unpredictable, avoid excessive spacing of the nails. The following is a suggested maximum spacing of nails for different thicknesses of plywood:

For 1/16 plywood use ¼ x 20 gage nails spaced ½" apart.
For 3/32 plywood use ⅜ x 20 gage nails spaced ¾" apart.
For ⅛ plywood use ½ x 20 gage nails spaced ¾" apart.
For ¼ plywood use 1 x 18 gage nails spaced 1" apart.

Normally, a minimum of one nail per wood member is required in gusset work such as is found in the construction of a truss type wing rib. It may be possible to use two or more nails in each member but it is not a good idea to space them closer than ¼" apart.

Corner blocks are almost impossible to clamp in place due to their awkward locations. Here is a place where a couple of nails will do a fine job of holding the blocks in position, and in assuring the desired clamping effect.

Another effective use of nails is in the attachment of ribs to the spars. Established practice dictates that ribs be glued and nailed to the spar by nailing thru the rib uprights at the spar openings. Nailing of ribs to spars thru the cap strip is normally avoided as this tends to weaken the rib.

Use of nails in attaching the plywood webs to box spars isn't too popular a practice now but, if you do nail your webs, it is advisable that you avoid driving nails too close to the flange edges. Stagger the nail rows slightly but keep the nails at least ¼" from the edges.

As in any method of applying pressure to a glued joint, if the joint is visible, the indication of an adequate joint is the appearance of excess glue squeezed out in a continuous seam at the glue line. Any portion of the joint which does not display this evidence should be subjected to close scrutiny. Possibly additional pressure by clamping or thru the closer spacing of nails may be necessary. For joints that are hidden from view, you have no guidance except to follow accepted good practices which will normally yield good results. Tapping the structure lightly with a hammer along the invisible glue line, should result in a solid sound to your ears. An odd or hollow sound conversely, could be an indication of a poor glue joint.

Develop the habit of marking the internal structural outlines on the outer skin to serve as a guide to nailing, and for future reference.

It is just a little thing, but always remove and replace any bent nails . . . do quality work. Drive the nails in snugly but do not bury the heads deeply into the plywood fibers.

SOMETIMES NAILS CAN CREATE PROBLEMS

When you do use permanently driven nails in the assembly of your glued wood components, all sorts of annoying results seem to crop us as construction proceeds and the smaller parts are progressively mated into larger assemblies.

Take the situation where you have just assembled an elevator and you must bevel and fair the trailing edge gussets in preparation for the fabric covering. The quickest way is to use a sharp knife or a block plane but . . . Yipes! Nails! Never happen, you say? Well, even if you carefully plan not to drive the nails too close to the areas to be later beveled . . . after you finish your beveling exercise look at your plane or knife blade. It will probably look more like a comb that a cutting blade.

You might as well take time out and remove all of the offending nails even though *that* may not be the easiest of tasks.

It is better to be prudent and do the gusset beveling job safely with a file or disc sander as either will zip right thru any overlooked nails. A file does a nice job and it can't be beat for final finishing, but it is a good bit slower than the disc sander.

You might run into another irritating situation. Holes for a fitting need to be drilled in the vicinity of a gussetted joint. You mark the spot and start. As luck would have it, you strike a nail and the drill wanders off center, enough at any rate, to affect the fitting's ultimate alignment. This is really annoying for once a hole is drilled, you are stuck with it.

(Figure 2) Photo of Corner Block Nailing Jig

Most of the classic arguments against the use of nails in wood structures are essentially statements of petty annoyances and may be summarized as follows:

1. Nails tend to work out in time and can cause ugly little localized bumps in the covering or finish of the aircraft.
2. "Iron" nails, due to the effects of the natural moisture content of the wood, can rust and can cause messy surface stains and streaks. (Of course, brass nails don't rust or stain.)
3. It also follows that three or four pounds of nails used in the structure would cause your bird to weigh three or four pounds more than if no nails were used.

(Figure 3)

Still another point to weigh. Even though using nails appears to be an easy assembly technique to use in gluing operations, this nailing business is a time consuming and tedious process.

Among the sophisticates, it is generally accepted that nails contribute little if any strength to a joint. Actually, they do no useful work at all once the glue has dried. Hence, it may be said that the nails are just so much dead weight and are going along for a free ride. Admittedly, however, it is mighty hard to argue down that "Steely-eyed" craftsman of Ye Olde Casein Glue Days, who insists that gussets glued and nailed look mighty formidable and give a guy that nice inner feeling of security. If using nails in your gluing operations does this for you . . . go right ahead. You will have plenty of company.

On the other hand, if you are going all out for weight control and appearance, and do own a dozen or two of assorted clamps, by all means clamp all of your stuff. Use as few nails as possible and save on both weight and money.

SO! WHY USE NAILS? Well, in addition to that vague feeling of security already mentioned, strong arguments for the continued use of the humble little aircraft nail do exist.

One such example is in the construction of a truss type wing rib. Should you decide to use no nails to hold the gussets in place, the gussets must then be clamped, right? If clamps are used, they must remain in place for the minimum of several hours and preferably overnight. To keep from marring the wood, it is necessary to prepare a clamping board or individual back-up blocks to prevent the metal clamps from sinking into your work and messing things up by crushing the wood fibers. Besides all of that, you then find that you have used up most or all of your supply of clamps and that your jig is not available for assembly of another rib. Employing this method your production will be limited to perhaps one rib per day with gussets still remaining to be added to the opposite side later, when once again you must go thru the same process. As if that isn't enough, you have to mix glue many more times to construct a given number of ribs, each time some is left over and wasted. A side result of this is that your glue costs will be higher than anticipated. By this time the temptation to use nails becomes overwhelming when you realize that . . . in contrast by using the glue and permanent nailing method, no clamps are needed and the rib may be removed immediately from the jig and another inserted for quick uninterrupted production and increased glue economy.

Of course, there are some places where clamps can't reach and you realize that a nail is just what you need . . . just one tiny little nail.

NAIL STRIP GLUING

When skinning wings and when attaching plywood skins to larger structural assemblies, you may prefer not to use permanent nailing and yet you find it impossible to apply pressure with clamps as both sides of the structure are not accessible to clamps. In this case you will find it handy to use temporary nailing strips.

A nailing strip is used to impose sufficient pressure along a glue line to assure a good strong bonding of the wood parts. Once the glue has dried the nail strip, having served its purpose is carefully pried loose and discarded. This leaves a nice smoothly skinned surface unmarred by nailheads. You can also use the nail strips any place where permanently driven nails would be objectionable or not desired.

I prefer to use nail strips made of scrap white pine although plywood or any common wood 1/16" to 3/16" thick can be used. The most usual widths are ⅜" and ½" but in some cases may be as wide as 1¼". Actually the width of the nailing strip should approximate the width of the underlying member. If required, two or more strips can be nailed down side by side to effectively spread the pressure over a larger area. Make your strips as long as necessary to span the area of intended use.

Prepare the strips in advance by pre-nailing a batch of the battens. Remember the nail must be long enough to pass thru the strip, plywood, and then into the structure.

As the strips will be carefully split away and nails removed, it is not a requirement that the nails used in nailing strips be of aircraft quality. Any nail of similar size will serve the purpose. Nail spacing in the strips is governed by the thickness and rigidity of the strip. As a rule spacing of nails should not exceed 1½".

Wherever nail strips might come in contact with glue, it would be wise to use wax paper under the strips to prevent them from being glued inadvertently to the structure.

Where there is a sharp curvature of the surfaces, it is better to make your nailing strips of plywood (short

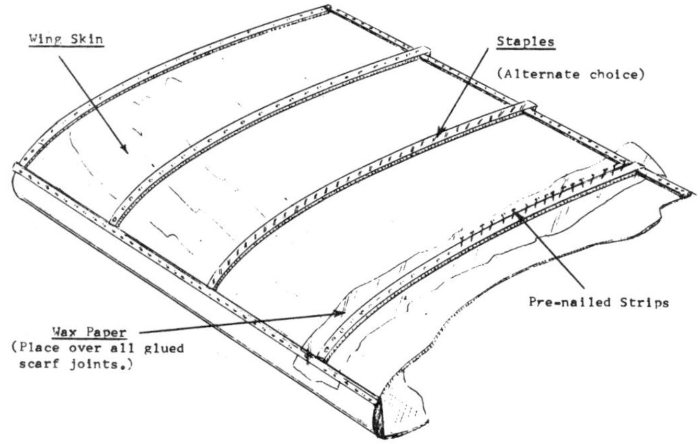

(Figure 4) Typical Use of Nailing Strips

grained) or use thinner, more flexible pieces of solid wood. This flexibility will permit them to more easily conform to the contour of the surface to which they are being applied.

Nail strip clamping is not normally recommended when the plywood's thickness exceeds 3/16". Instead, use clamps or permanent nailing if necessary.

In nailing down the strips always work from the center of a panel section outwards. The first strip is usually nailed down along the primary structural support. The subsequent strips are then worked and fastened down outwards in all directions progressively to insure that you do not wind up in one end with bulges or gaps in the joint which have no place to go. This sequence of fastening will ordinarily insure good results providing that the surface being covered is free from compound curvature.

NAIL STRIP REMOVAL

As you may have suspected, the removal of the nailing strips after the glue has dried, can be a bit of a problem to say nothing of the valuable time which is consumed. Whatever technique you may devise, take care not to damage the plywood skin or to excessively enlarge the nail (or staple) holes as the strips are pried loose. Sometimes it is easier to carefully split the wood strips away from the nails with a sharp wood chisel, clear the mess away and pull out the nails with a pair of diagonals.

Nail holes remaining after nailing strips have been removed may be filled with a wood filler and surfaced before the final finishing of the surface. Filling the tiny nail holes may not be worth the effort if the plywood surfaces are to be later covered with fiber-glass or fabric. Just sand the surface to remove roughness around the nail holes.

Let's sum things up by suggesting that you can use nails any place you want to providing you remember that the use of nails really has nothing to do with the ultimate strength of a glued joint. As a matter of fact nails must never be used to carry structural loads. (It's in the book!)

Plywood door skins, or similar material, must never be substituted where A-A Marine or Aircraft Ply is specified on the plans, especially on the main bulkheads. Some dealers have passed off this material as A-A Marine, so you must check it yourself. Marine and aircraft plywood will have outer plies almost as thick as the core, while door skins, etc., will be practically all core with very thin outer plies in addition to other deficiencies. Accordingly, where the design is based on the outer plies carrying the load in the direction of the grain, as in the main bulkheads, the necessary material just isn't there. Proper marine plywood should have "A-A Marine" branded into one edge of the full sheet.

W. S. Evans, EVANS AIRCRAFT

Very few amateur built aircraft are constructed without some wood being used. Most composites use wood as the basic frame in their construction. Many designs, including certificated aircraft, with metal fuselages are constructed with wood wings. And with few exceptions, tube and fabric aircraft have wood wings. Note plywood skins on lower surface of wings of this Wittman Tailwind set up for final assembly and FAA inspection. Top wing skins will be glued on following sign off by the inspector.

Building the Isaacs Spitfire
(A Homebuilt Spitfire)

by John O. Isaacs (EAA 41920 Lifetime)

(Photo by A. J. Clark)
John Isaac's Sport Spitfire.

ORIGINS

Upon completion of my 65 hp 7/10 scale Hawker Fury I biplane in August 1963, it was suggested to me that I should build a baby Spitfire. With my kind of background as related in Part I of this story, the idea was interesting although the design and construction of a clean cantilever monoplane presented a formidable challenge after the simple boxlike structures of the wooden biplanes of my initial experience. Also the account of my Fury published in *SPORT AVIATION* (May 1964) generated a fair amount of interest leading to a sizeable and time-consuming worldwide mailbag. But it was mainly the indigenous encouragement which eventually led me to double the power and generally update the Fury to Mark II standard (see Fury Finalized, *SPORT AVIATION*, December 1968). Despite the intervention of this additional work, I found myself carrying out feasibility studies and producing 3-view G.A. drawings of the Spitfire as early as 1964.

By the summer of 1968 the Fury II version had been finalized, proven and sold to allow me to devote all my energies to the Spitfire project and it was only in that same year that, after much heart-searching, the final and irrevocable decision was made to go for 6/10 size even though it would mean some deviation from scale in the cockpit area. A cunning combination of different features from various Marks enabled me to produce a hybrid version while retaining a true, classic Spitfire flavor.

Early in the game I had received assurances from my old friend, Ray Hilborne, that he would kindly stress the airplane for me just as he had done for the Fury; it was a weight off my mind. Naturally, we decided to work to the British Civil Airworthiness Authorities requirement of +9 −4½ G (fully factored) so that aerobatics could safely be carried out.

Major decisions had to be made concerning the wing construction and the undercarriage. It would have been convenient to build the wing in three sections but deailed investigations confirmed the expected weight enalty and complication of steel connecting fittings and ozens of bolts to transfer bearing loads back into the imber. Thus the simplest solution lay in a one-piece hrough wing and I opted for a traditional two-spar ayout rather than the single-spar French Jodel sys-em since the Spitfire wing is exceedingly thin out owards the tip, a fact which dictated the overall ply kin covering to meet torsional stiffness requirements. Vith the benefit of hindsight it is generally conceded oday that the Spitfire's undercarriage track (tread) vas too narrow and I was determined to improve the round handling of my airplane by widening it. Since, on he real thing, the legs folded outwards my wheels would ave had to fit into an even thinner part of the outer ving so it seemed logical to fold inwards, if at all. While tried to make provision for this, it proved difficult on a ery small airplane to achieve successful retraction so he final fixed gear decision can be blamed on the Bill Stout dictum of "simplicate and add more lightness". The same philosophy was responsible for my not fitting railing edge flaps to the wing. The validity of such udgments, often inevitably influenced by personal ex-eriences, necessarily had to await flight trials on the ompletion of the airplane years later.

Starboard wing tip skin about to be glued down.

THE MAINSPAR

By Easter 1969 I had a fair idea where I was going nd as there was a small amount of spruce left over from he Fury I had this sliced up to enable me to make a nodest start, traditionally, on the tail surfaces while thought about the wing mainspar. Frankly, this fright-ned me as it was to be a 22 ft. long, built-up, curve-apered box and I was short on experience of this kind f thing (although I had read articles on "how to do" ometimes written, I suspect, but those who had not). The spar was the biggest single piece of airplane need-d to get the whole project under way and I have to admit hat, to save time, consideration was given to having t made by a professional sailplane manufacturer, but hey wanted too much money for it and an additional 00 pounds sterling for the jig alone.

So it was at this critical moment that I became as-ociated with Leslie Broomfield who had asked me to ct as inspector on the Taylor Monoplane he was build-ng. We had arranged to do some economic joint order-ng of materials for our airplanes and I was fortunate ndeed to get Leslie interested in my spar problem. It oon transpired that he was a trained and highly skilled voodworker with his own well-equipped workshop. To-gether we selected two sturdy baulks of spruce which then had ripped up into estimated required sizes, in-luding a goodly number of 5/16" thick planks for lami-ating the spar booms. Together we screwed $10 worth f 6" by 1" wooden floor boards to crossbattens to form a lat base 24 ft. by 2 ft., which we laid on the floor of the vorkshop classroom of a local school during the summer olidays. I drew out the spar shape in pencil on the fresh vhite timber, the 6^0 dihedral originating either side of n approximately 3 ft. wide straight center portion. At bout 18" intervals we screwed wooden blocks outside he spar contour and to these, in turn, we screwed a long edwood plank, on edge, set flush with the penciled urved contour. By borrowing every available G clamp nd a few sash clamps which totaled 3 or 4 dozen we were ble to glue and clamp the top boom laminates to the edwood former and leave them overnight to set. The op (compression) boom has seven laminates at the root which are progressively shortened so that at the tip only he outer one remains.

Throughout the aircraft I used Aerolite 306 which s a urea formaldehyde powder resin requiring a water

Starboard wing tip and rib detail.

Leading edge skins in place.

mix and this is "cured" upon contact with a formic acid solution obtainable with variable setting times. As a precaution against the spar sticking to the jig thin polythene sheet was interposed and retained in place by staple tacks. We had to work fast on a very hot summer afternoon to get the seven slippery planks firmly fixed in the right positions before "curing" set in after about 20 minutes maximum. Another hazard, only apparent when all the parts of the laminate sandwich were glue-smeared prior to assembly, was that the premixed 2 lbs. of glue was barely sufficient. A day or so later the basic top boom was removed from the jig and the wooden blocks and redwood plank repositioned on the base to enable us to laminate the four planks which comprise the bottom (tension) boom in the same fashion.

Some work was then necessary to clean up and cut the 22 ft. long booms nearer to finished shape. Taking advantage of our temporary surroundings the booms were, on occasion, clamped in the vices of two or three woodwork benches at the same time to avoid a whippy overhang while they were worked on. Since the full-wing-depth spar is of optimum design, steps caused by the progressively shorter laminates had to be planed away to produce a smoothly tapered inside surface thus avoiding undesirable stress concentrations due to sudden changes of section. The width of the spar in plan view also tapers from over 4" at the center to about 1" or less at the extreme tip so surplus timber was now cut off roughly to shape. The front face of each boom had to be planed truly flat and level ready to receive the 3/16" thick forward ply web which extends the full span.

The last spar jigging stage was to remove the redwood plank and rearrange the wooden blocks on the jig base to locate both booms in their correct relative positions, forward face uppermost, while spruce blocks were cut, fitted and glued between them at rib, undercarriage, and wing attachment stations. Finally the forward ply shear web was glued on to join both booms, scarfed joints being necessary to make up the full 22 ft. span. The basic mainspar had been assembled only three weeks after we started to make the jig although much careful detailed shaping, drilling and attachment of the aft web still remained. With the end of the school holidays looming up we hastily dismantled the jig and evacuated the embryonic spar for storage in Leslie's long garage while I anxiously awaited planning permission from the local authorities to have a 24 ft. by 10 ft. shed built in my back garden. This was put up in December, but while waiting, I made the simple truss-type wing ribs, each of which is in three separate pieces due to both spars being full wing depth. Each rib is also of different chord, different contour and different thickness/chord ratio but all are scaled down from the NACA 2200 series airfoil shapes used on the original aircraft. Each rib was drawn full size and nails hammered into a plank to jig them to contour. As they were completed the chord and front spar datum lines were carefully penciled on each. The lighter rear spar was built in three separate pieces too, so that the forward-cranked outer portions could be scarfed to the center section during wing assembly.

Sporadic work continued on the main spar until, with the aft web (which fades out at rib 6) glued on, the completed unit was set up under Leslie's pillar drill to drill the ten holes each side which secure the cantilever undercarriage leg attachment socket. We also pre-drilled the wing-to-fuselage attachment holes. With great interest I weighed the completed spar to find that it clocked in at a little over 50 lbs. The specter of the main spar construction was finally laid to rest.

In retrospect I consider the job a formidable undertaking. Some really hefty chunks of timber are involved

The author showing the leading edge top skins ready for application. Notice that the areas where glue will be applied are left unvarnished. The dark strips between rib glue lines are ply stiffeners.

Aileron bellcrank and hinge details at inboard end of wing.

Partly finished aileron placed in position on wing for trial fit. Rib detail is obvious here.

ıd the puny tools which had proved suitable for whittling ıe small, delicate scantlings of, say the Fury were uite inadequate. I had the tremendous advantage of eslie's woodworking knowledge behind me, the use of is extensive variety of cutting tools, his sheer physical nergy applied when pushing a two foot long trying plane long the booms to get them level, the temporary spaıous facilities and power tools of a school workshop. he less fortunate could get there in the end but the job ould take longer.

THE WING

In the spring the main spar was transported to my wn new workshop. The old spar-jig base was reassembled ıd laid on trestles to form a table 2 ft. wide. I did not onsider it worthwhile or necessary to level this up truly ıroughout its length and breadth since it was my intenon to utilize the completed spar itself as the key component to which all else could be set. To this end a penled datum line was drawn the length of the spar. It as, of course, imperative to correctly build the incience and symmetry into the wing to achieve good flight naracteristics since stiff cantilever wings cannot be rigged" by slight warping as is sometimes possible on iplanes.

The main spar was blocked up a few inches above the ıble top and in flying attitude. The center portion of he rear spar was set up in its correct relative position nd the two spars joined together by bolting in the two teel tube tie bars which transmit undercarriage loads om front to rear spar. The center portions of the inoard ribs were then carefully fitted and glued and crewed between the spars.

Next Leslie was called in to scarf the outer portions f the rear spar to the center section while I ensured ıe correct geometrical position of the outer tip by checkıg from my penciled lines that dihedral was maintained ombined with an upward tilt caused by the "washout".

Throughout, my technique was to lay a wooden raightedge across the jig table (chordwise) below the par at a rib station. This was then packed as needed ntil horizontal as indicated by a spirit level. All I had do then was to ensure that a vertical dimension from ıe straightedge up to the pencil line on the front spar as identical to a similar dimension at the rear spar. enciled lines on the ribs were similarly used to set them ertically in relation to the spars. In this manner, using ıe penciled front spar datum lines previously marked n ribs the linear twist from $+2^0$ to -0^0 30' incidence as achieved automatically and simply as the straightdge was moved outboard from rib station to rib station nd packed level. The center portions of all ribs were ssembled first throughout the full span thus constructıg a sort of ladder, the stiles formed by the spars, the ibs representing the rungs. To this firm foundation nose ib and trailing ribs were individually fitted and seured working from the mobile straightedge.

When the nose ribs were in place a continuous spruce ading edge member was pulled round over each and lued into position to act as a butt-strap for top and ottom ply skins. Later, a narrower spruce strip was lued to the front face of this and the changing LE radius arefully shaped and checked by card templates made or each rib. The completed structure was inverted onto uitable supports built up on top of the jig table while the ½ mm ply bottom skins were glued on over the whole urface after which I had a robust, rigid wing which ould be safely manhandled the shape now having been ermanently established. It should be remembered ıat, when the wing was right side up it could be supported on the table at the center and steadied with light, aised, wing tip supports but when inverted the dihedral

Assembled wing with bottom skin fitted. Note redundant diagonal brace.

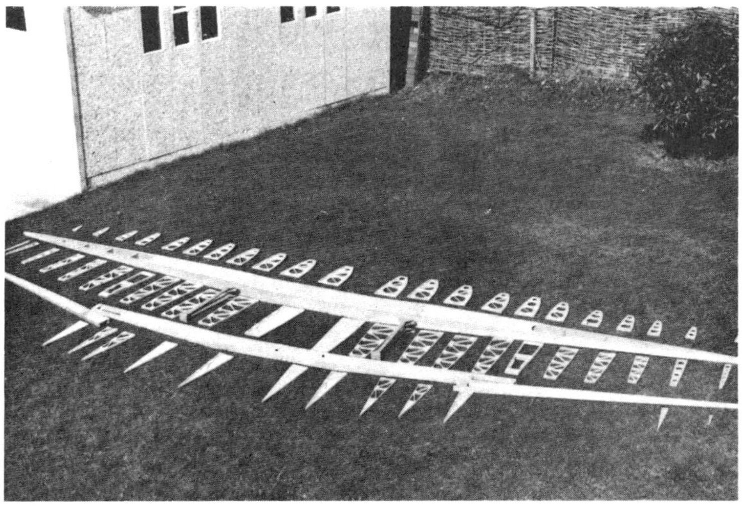

Sport Spitfire wing prior to assembly.

Main spar and rear spar center portion in Leslie's garden.

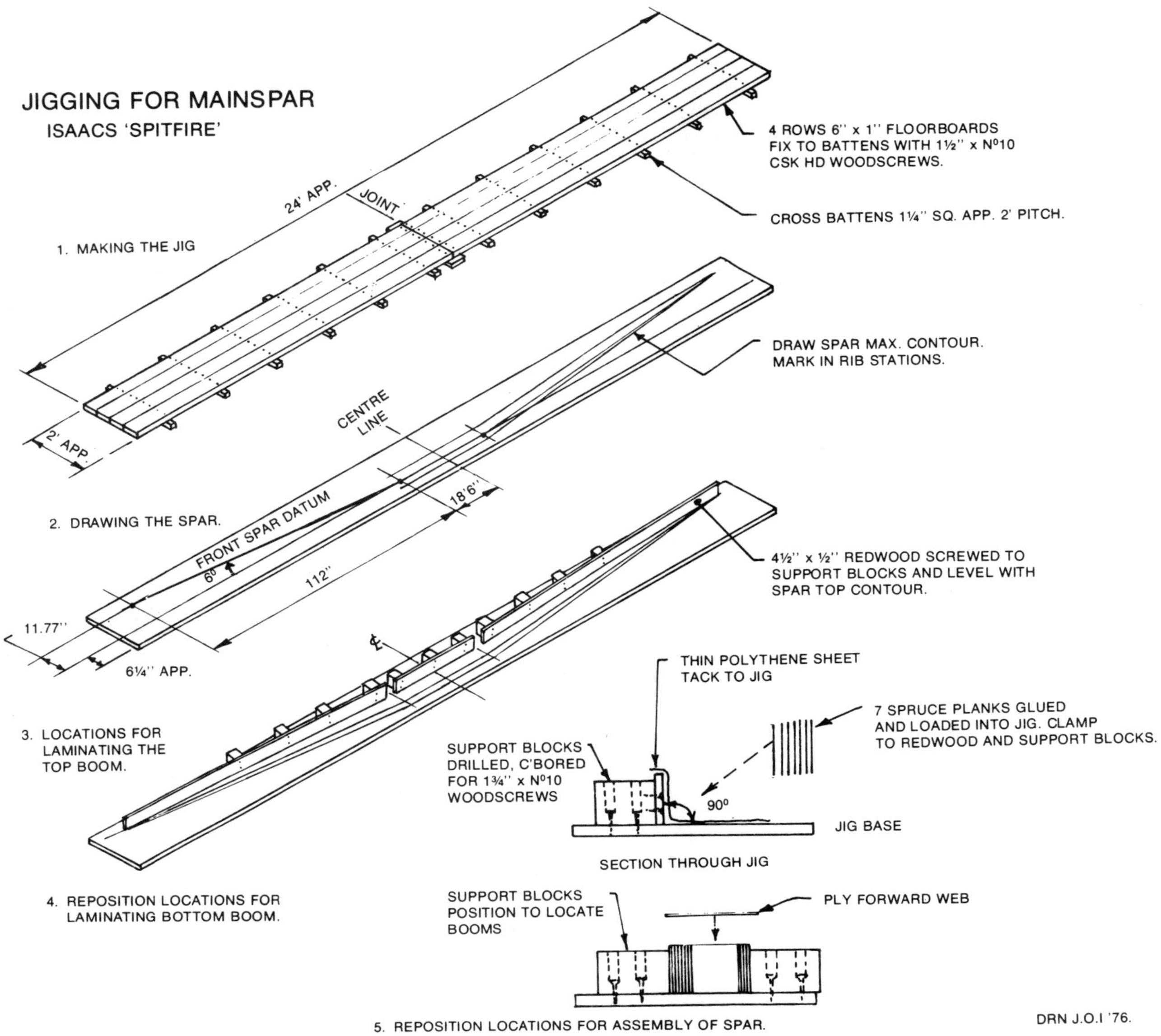

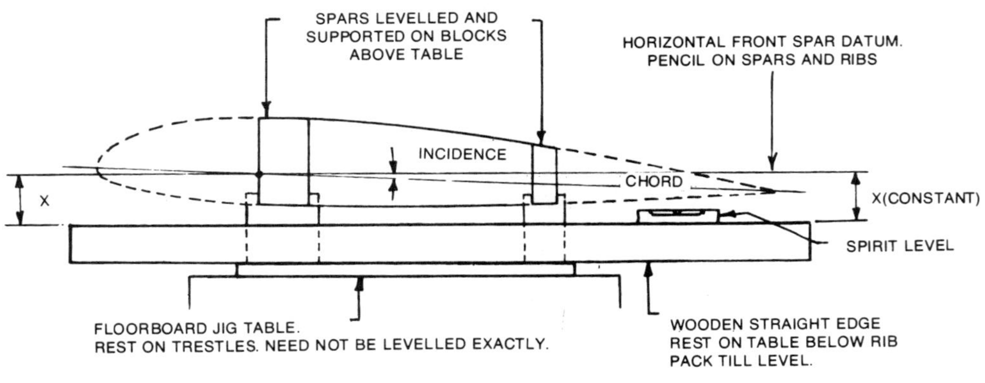

BASIC METHOD OF SETTING RIBS FOR GLUING AND SCREWING TO SPARS.

Horizontal stabilizer detail.

Main spar top boom laminates glued and clamped in jig.

raised the center of the wing some 18" above the table while the tips were almost touching it. Later, when the table was removed and the wing was supported only on trestles at the center, I evolved a method whereby I could turn it over single-handed despite its 22 ft. length and 5 ft. root chord.

The bottom skin covering is fairly flat but there is considerable, apparent double, curvature on the top LE skin. Nevertheless the ply panels were successfully applied and firmly secured to the butt-strap by staples and brass brads while the glue set. Final LE contour shaping and checking by template was then possible.

Initially, I hoped to save some weight by having a fabric trailing portion of the wing so I built in diagonal drag bracing members on the prototype and these appear in the photographs. They are not now considered necessary as the wing skin coverings can carry the drag shear loads; on the prototype they do help stiffen the ply by reducing the size of the unsupported panels.

In my own time I could now install and prove the aileron control run before fitting the top covering. All controls are operated by the light and simple method of stranded cables with only a short final rod from bell crank to control surface operating lever. The ailerons are of the Frise type, patented by the British designer Leslie Frise of Bristol circa 1922 and whose name has so frequently been misspelt by ill-informed writers. This simply means the hinges are inset so that upward movement of the aileron causes its leading edge to project below the bottom contour of the wing but downward movement is possible without producing any such projection or gap and it results, of course, in increased drag on the downgoing wing on the inside of a turn. In addition to this feature I incorporated a differential system and fitted mass balance weights against flutter. Further, I increased the inboard area slightly larger than scale in an attempt to improve what I had always been told, was the least pleasant of the Spitfire's controls.

To prove the full aileron circuit I had to temporarily fit the control column, its bearings and the articulating rod. This unit being mounted between front and rear spars it could only be finally assembled after the wing spars were slid up into the slots left for them in the bottom of the fuselage. Since I had not clearly decided what to do about the aileron shroud and because I wished to build and fit the ailerons while the inside of the wing was still accessible it was to be some considerable time before the final top skins were assembled. Meanwhile, the bulk of the work done, it could be stored leading edge down against one wall of my narrow shed.

So, again in retrospect, the difficult curved wing had proved really quite easy to build with a minimum of equipment provided one was prepared, as always, to go about the tasks patiently and methodically. At about this time, in April 1971, I received a tremendous boost to morale when President Paul, on a visit to England, walked into my humble workshop, enthused over the wing and said in his quiet, shrewd way, "It's the right size". I felt certain that my early crucial decision over choice of scale had been vindicated.

THE FUSELAGE

Now able to turn my attention to the fuselage I first lofted it to a reduced scale on the drawing board and, working from the scaled down real Spitfire cross-sections, I widened them slightly in the cockpit area, gradually reducing the increase in width so that they tapered back to almost exactly 6/10 scale again at the stern and at the nose. From this layout I worked up a full size drawing showing a nest of fuselage frame outlines. A print was pinned to a flat table of about 3 ft. by 2½ ft., again made of floorboards, and covered with transparent polythene sheet. By locating the various members with removable blocks or nails each frame was assembled in turn on the same jig base. The slightly curved frame sides were cut from flat spruce planks while the rounded tops and bottoms had to be prelaminated from thin spruce strips on small separate jigs. They were joined with ply gussets and spruce corner blocks.

Forward fuselage structure.

The first assembly problem was how to ensure that the wing would match the fuselage correctly; with a 10 ft. wide workshop it was clearly impracticable to build the whole fuselage "in situ" on the wing. After some thought I decided to start by making the forward part of the fuselage actually on the wing, and across the shop, leaving short lengths of longeron protruding aft of the wing for scarfing on the rear parts later. Thus the four wing attachment holes were drilled in the fuselage frames which were then bolted to the wing while I built up the forward fuselage structure from the fireproof bulkhead frame back to the cockpit. The completed rigid framework was disconnected from the wing which was again stored against the shed wall.

Since the top longerons are the horizontal rigging members it was preferable to build the fuselage upside down, locating the top longerons, and to that end the next stage was to make a simple assembly jig to secure the straightness of the whole fuselage. Some of the old original wing jigging floorboards were used to build a sort of platform round the outside of the fuselage contour at the top longeron level.

A string was drawn tightly from nose to tail to act as a centerline while frame stations were marked out in pencil at right angles to it on the platform planks. A square-section wooden crossmember was screwed to bridge the platform at each frame position and to which the inverted frames could be clamped, athwartships alignment being achieved by matching frame and jig centerlines.

The forward fuselage structure was placed upside down into this jig and the aft portion of the (inverted) top longerons scarfed on. The rear frames were next set up vertically in the jig and fixed to the top longerons. The tricky job of scarfing the aft bottom longerons to the forward stubs was accomplished and, when set, they were pulled down and into the notches in the frames. After carefully shaping and fairing the top and bottom longerons they were joined by applying the 2mm side skin panels so that the now rigid fuselage could be removed from its jig for the assembly of the curved top and bottom ply decking. Before gluing on the turtledeck I temporarily installed a dummy rear spar on which I could mount the cockpit elevator lever so that the control cable run could be proven. Cable routing from the pendant type rudder pedals was also checked.

The fuselage front top decking, complete with attached windscreen, is quickly detachable which gives easy access to both the cockpit floor area with its rudder pedals, brakes and hydraulic lines, tank installation and also to the back of the instrument panel. This desirable feature carried a penalty in that a single diagonal tubular strut had to be fitted across the open fuselage at top longeron level to carry engine torque loads. On the real Spitfire this problem was met by using two tubes forming a plan vee but, being a larger airplane, there were two tanks, one above and one below the bracing. On my machine the single tank has a tube passing through it to allow the clear passage of the structural brace tube as is common on some Piper wing tanks. Even so the welded aluminum tank retains a fuel capacity of 10 Imperial (12 U. S.) gallons.

Fuselage jig. Note stored wing.

FIRST ASSEMBLY

Despite my careful jigging endeavors to guarantee the symmetry and correct rigging of the finished airplane I just had to remove that nagging doubt by arranging for some robust friends to carry all the parts into the back garden one evening for a trial assembly. So for a brief hour the fuselage was squeezed between two garden walls, the wing jammed between an apple tree and raspberry canes while I ascertained that the diagonal symmetry between wing tip and sternpost was within a most satisfying ⅛".

For some time I had been deliberating the problem of wing fillets for the manufacture of which it would have been desirable to have wing and fuselage assembled. As the size of my workshop prevented this, I got down to making a rough dummy wing center section to hang on the fuselage in the hope that it would help guarantee that the large fuselage-attached fillet would reasonably match the wing contour. Although I felt that the fillet could well have been made of some form of expanded polystyrene-type plastic, I stuck to the devil I knew and, having lofted the rather tricky contours, found that I could build it on outside the basic fuselage skin using light formers and curved plywood panels. For ease of construction, the bottom of the fuselage is fairly flat over most of the wing chord back to the trailing edge after which it assumes a curved shape, the transition being achieved by a double-faced frame. The resultant "step" is afterwards hidden when the lower fillet panel is glued in place.

THE UNDERCARRIAGE

Two uncertain areas of the aircraft now had to be finalized; one was the undercarriage, the other was the canopy.

I studied "Landing Gear Design" by H. G. Conway and decided that rubbers in compression seemed to give

Fuselage upside down in assembly jig.

the most efficient compromise and I knew from my personal Chipmunk flying how reliably well they worked. At the local rubber manufacturers I was given a handful of rubber samples of varying hardness, but there seemed to be little useful technical data to work on so Ray Hilborne and I spent an interesting day at the University carrying out compression loading tests on a machine, fortuitously free during the summer vacation. From our findings we drew graphs, selected the most suitable sample, made intelligent assumptions to be used for stressing and detail design. A good friend and colleague, Tony Clark of British Olympic pistol shooting renown, machined a mould and I ordered the rubbers while I set about detail design of the cantilever leg and sought out the required steel tubing. I bought an expensive chunk of phosphor bronze to make the bushes for the telescopic leg which Tony kindly machined for me. The torque links were fitted, with further reference to Conway, on the front face of the leg. A few weeks later I acquired a pair of 5.00 by 5 Dunlop wheels and brakes as used on the Beagle Pup. I was now able to fit the assembled undercarriage leg units into my wing-mounted sockets and check them for symmetry and forward rake, an anxious concept since the relationship of the wheel contact points with the ground to the center of gravity of the finished aircraft would be of vital importance and difficult to alter if not correct first time.

Fuselage with plywood installed on sides. Note the slots for the wing spars.

THE CANOPY

The canopy problem was twofold. First, how should I design it; should it hinge sideways — perhaps the easiest solution — or should it slide back as on the original and, secondly, how should I make it?

Regretfully, I abandoned hopes of using a standard sailplane canopy so I lofted the shape I required and checked it by sitting in the cockpit with wooden battens and card templates over my head. Since, with a typical narrow fighter canopy of about 12 or 14 inches base width, access was impossible, the answer lay in attaching the canopy to part of the fuselage sides so that both these and the canopy would slide aft allowing an adequate entry width of 18" across the top of the shoulders.

Seeking expert advice I bought some Jelutong, a wood I had never heard of, as it was said to be easily shaped. I had this cut into lengths of 2" by 1" thick sections. I drew full size level line sections onto the lengths of timber and carved them to shape. The hollow mould was then built up by gluing together the Jelutong laminates with cabinet makers Evo-Stik rather than the glass-hard-setting Aerolite. Shaping, frequently checked by template, did prove easy enough although it took the equivalent of nearly three weeks fulltime work to finalize two moulds, one for the fixed windscreen and one for the sliding canopy.

The general view was that it should be possible to drape mould the ⅛" perspex I intended to use but, due to the uncertainties of the weekend availability of an industrial oven and a team of helpers, I eventually took the moulds to the professional firm of Suntex who make canopies for the industry. Acting on their advice, I modified my mould slightly and marked on it the estimated required cut line which could only safely be established by additional laborious jigging on the fuselage.

PER ARDUA . . .

Meanwhile the sad though timely demise of a Jodel, condemned through glue deterioration, resulted in my acquisition of its half-lifed 100 hp Continental 0-200 and its metal ground adjustable French Ratier propeller. Furthermore, the engine was American built and, dare

John Isaac's Sport Spitfire

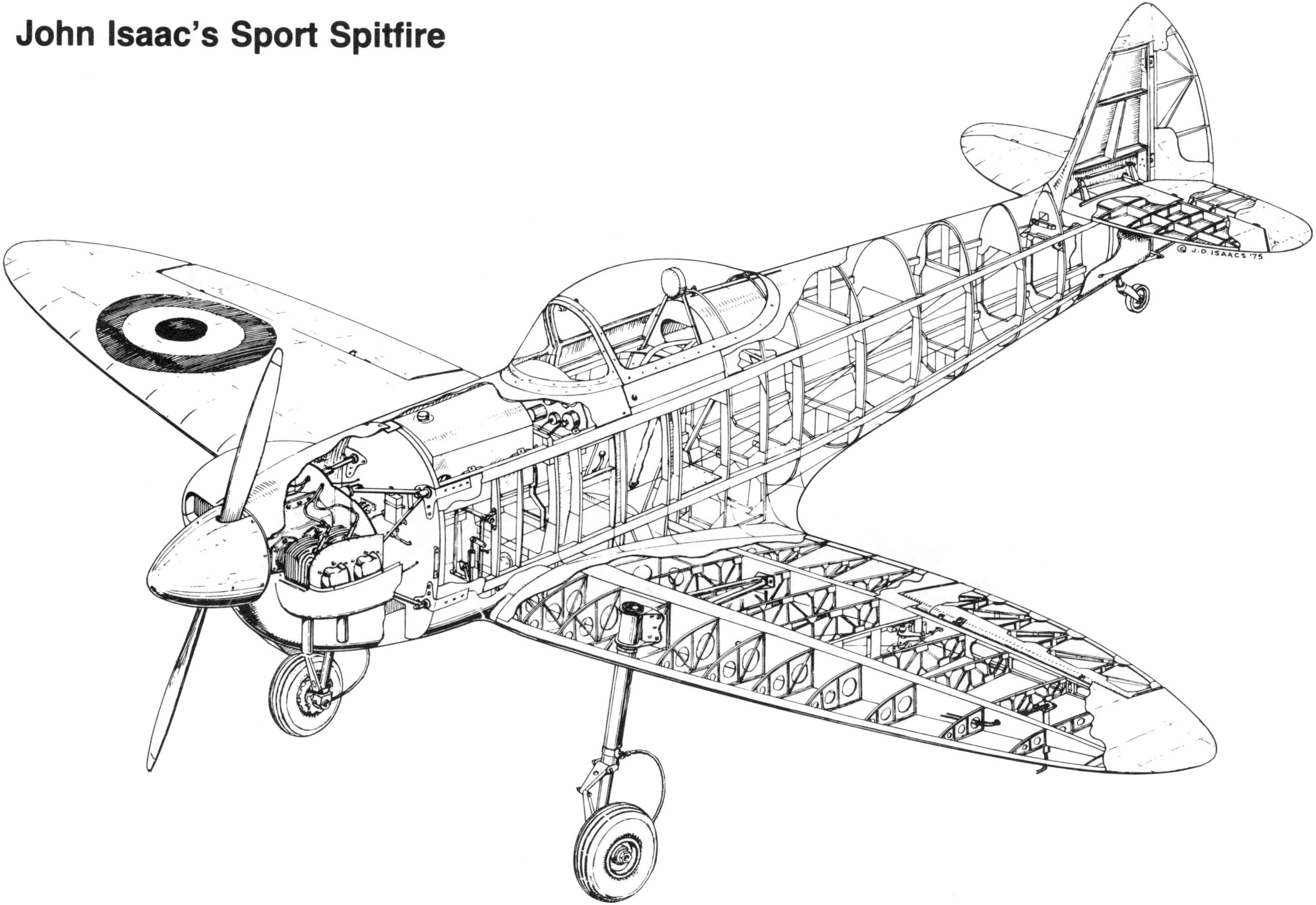

Wooden canopy moulds ready for drape moulding perspex over them.

I admit it, they are said by some to be better than our license-built versions. As it was unlikely to be required for some while, we struggled up the high staircase with it to store it on the bedroom floor where it joined the almost complete tail surfaces and ailerons. Indeed, taking stock of the situation revealed the depressing fact that after four year's work not one single component part of the aircraft had been completed.

No matter how carefully applied, thin ply always seems to form hollows or, worse, even ripples between anchorage areas on ribs and frames. But I hoped to emulate the magnificent finish obtained on sailplanes so I went to see individualist Ralph Jones who repairs sailplanes and also excells at flying them. He freely initiated me into the technique of employing a German chemical mix of Laromin and Epicote into which phenolic micro-balloons were stirred before the whole pasty mess was slapped into the unwanted hollows to be smoothed by drawing a bridging steel straight-edge or rule across the surface of two adjacent ribs. After curing, the high spots of the brown mixture were removed with sandpaper to reveal where a further build up was necessary — a process repeated as often as needed.

After an initial abortive attempt using an alternative resin which failed to cure and had to be scraped off my tail surfaces, I reverted to that recommended by Ralph and success began to crown my efforts. Having tediously filled and rubbed practically the whole wetted surface of the airplane thin madapollam was stuck over the ply with clear low-tautening dope. As the successive layers of dope accumulated and were rubbed down with wet and dry abrasive paper for days on end, my fingertips actually bled.

Change of shape of the bottom of the fuselage before the wing fillets are built over it.

To take advantage of the summer weather, in August I decided to paint and Dudley Morgan kindly trailered over his compressor on loan for a week. My early paint schemes had been temporarily abandoned when I toted up the surcharges for small quantity orders and I settled for plain cellulose silver all over working to what is known over here as Scheme Z. It was cheap (relatively), it eliminated the need for an aluminum undercoat, and it would show off that beautiful elliptic shape. The last thing I wanted was to break up that superb contour with drab wartime camouflage schemes although had PRU blue been available I would have settled for that.

ENGINE COWLINGS

By the fall the engine bearers had been welded on a simple steel angle jig modified from the one I made for the Fury years before. The engine was temporarily installed in the fuselage then returned, with its bearers, into the house. Knowing the cowlings were likely to take long mid-winter weeks and having seen the sorry, wet looking engines which emerge from condensation-inducing plaster cocoons I evolved my own method. A wooden template replica of the fireproof bulkhead was made from the actual aircraft and bolted to the engine bearers with engine attached, propeller flange uppermost, so that I had, effectively, the nose of the aircraft in the house. Using my lofted nose lines I then built up a rough wooden framework around this to ensure clearances but in such a way that it could be "split" or dismantled and reassembled to the replica bulkhead without the engine inside. In this way the engine stayed in the warm while the wooden framework, reassembled on trestles in my workshop, became the basis of the "plug" onto which I fixed chickenwire netting to hold the usual plaster of paris. A plaster shortage hit the nation while halfway through this job, but I managed to obtain and use about 200 lbs. of best dental plaster. In January 1974 the Arab oil crisis hit the world and, apart from the fears of whether I would ever be allowed petrol to fly, all the oil-based plastics began to skyrocket so it was none too soon that I was able to enlist the help of Ron Rollison with his expertise and supply of vitally needed materials. Working at weekends and using 2 oz. and 1½ oz. fiberglass mat we first made female moulds from the plaster plug and then the cowlings themselves.

Completed fuselage. Detachable forward top decking not shown.

While we had all the materials to hand another problem resolved itself. For some time, thinking of aluminum sheet, I had been mentally wrestling with the problem of how to make the intricately curved windscreen and canopy external framing when Ron made the outrageous suggestion of using fiber-glass. I say outrageous because I was most apprehensive of the chemical's effects on

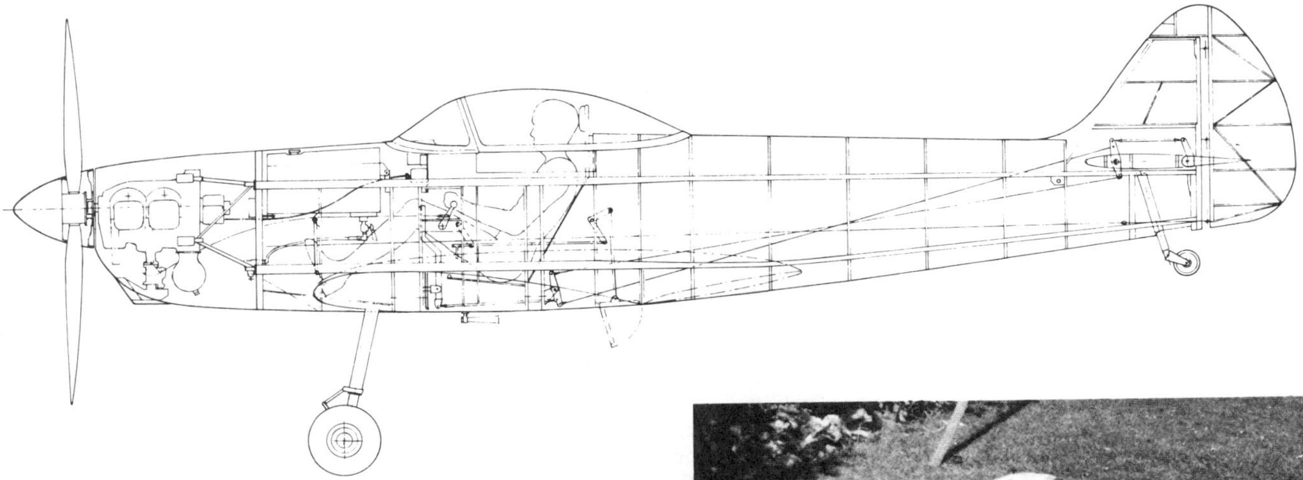

Dunlop wheel/brake unit, torque links and hydraulic fluid supply line.

Laminates for building the wooden canopy moulds.

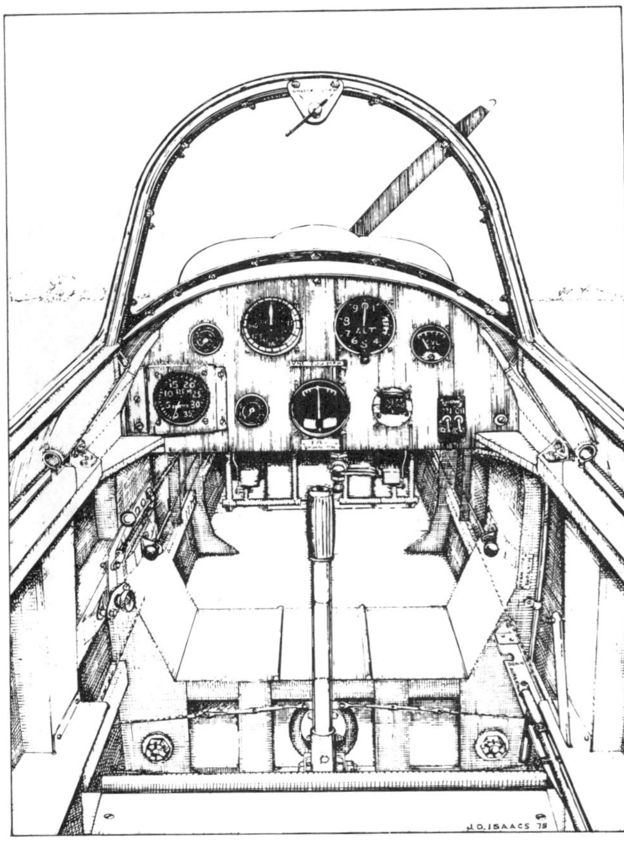

those hard won perspex components. Nevertheless, with a protective plastic sheet laid over the perspex this is just what we did. By the end of March I was able to break up and dispose of all traces of those messy plaster and plastic jobs.

The last remaining plaster and fiber-glass task, later in the year, was to make the spinner, the plaster plug being moulded by working a shaped board or template round a center point in the same way as the strickle makes use of the generating principle during loam moulding in an iron foundry. Now the engine could be reinstated and with it was opened up the exciting task of cutting and fitting the cowlings.

›ugh wooden structure before covering with plaster make engine cowling plug.

Parts of the master female fiber-glass cowling mould made from the plaster plug.

THE FINAL YEAR

With all the major components nearly complete, I ıd reached that interminable period of finalizing deıils. Wheel brakes bothered me since I had no experiıce in that field. I bought a pair of Cleveland master ylinders from Harvey Swack but it took a long time to ork out their modus operandi and the complete ciriit. When I built the Fury I used readily available parts ıch as the 1 lb. 3 oz. Tiger Moth type fuel filter, the vro Anson 11 oz. brass Ki-gas primer and large brass ıel unions but for the Spitfire I collected an American oz. filter, a 2½ oz. Kohler primer and light AN unions nd adaptors for fuel and hydraulic lines. In the event, o primer was needed since the Continental 0-200 had n accelerator pump. Engineer Dave Ring helped me ort out some of the engine hook-up details while I also uilt up VW mileage looking for parts.

It was during these later stages of construction that ell-timed *SPORT AVIATION* articles provided inırmed discourses on a variety of relevant areas. I read ony Bingelis on hydraulic brakes, Bingelis on engine owlings, Bingelis on pressure cowling baffles, Bingelis ı exhausts. Incidentally, as I have short external exaust stubs, I took my carb heat from an aluminum box uilt under the starboard rear cylinder and in conjuncon with the rear baffles.

In mid-August came disaster. As I opened up the rilled attachment holes in the canopy to accommodate ıbber grommets, as is standard practice, the perspex racked and a piece broke out of the forward edge. Later, hen recovered from the shock, I found that by cutting ff one inch from the front edge the damaged area could e eliminated and the negligibly shortened canopy ıved.

The imminent onset of winter traditionally signalng a busy period at my daily task hopes of moving the pare time aircraft to an airfield were postponed until ıe spring. I busied myself updating drawings, fussed ith details, tried to find a welcoming airfield, made ransport arrangements tentatively, and generally preared for the great exodus. Regrettably, the engine ould have to be removed to lighten the fuselage as oth it and the wing would have to be lifted over a five foot high garden wall and thence out to the street via a neighbor's sideway.

Mid-January provided a tonic during a thought provoking lecture I attended given by the great Spitfire test pilot, Jeffery Quill, who brought a recording of a voice from the past, R. J. Mitchell himself giving a radio talk on high speed flight.

FINAL ASSEMBLY

It was to be Thruxton again. On this same airfield where I had first put my Fury biplane together in that summer of 1963. But it was an uneasy Thruxton. So, on the Sunday before Easter 1975, the parts were safely loaded on a truck, delivered, unloaded, wing and fuselage joined, undercarriage assembled, and by evening for the first time ever my Spitfire was "on the gear". I spent the next fortnight's vacation working in a secluded side bay of the one large, bitterly cold hangar, while outside it often snowed. The unease increased until ugly rumor crystallized into the fact that all airplanes had to be out of the hangar by the end of April since it was to be rented out for agricultural storage purposes.

My predicament was that I had a completely unknown quantity which had to fly out, fast, or be thrown out. Could I make it in time? I applied to the Popular Flying Association's John Walker for an urgent Permit to Fly while working frantic weekends, deafened on one Sunday by 160 mph racing cars hurtling past the echoing hangar on the perimeter race track close outside.

By the middle of the month the Continental was running and it was a real thrill to taxi up the grassy slope to the pumps for fuel. The evening of Wednesday, April 30, saw an assorted collection of vehicles desperately removing the last of the unflyable hangar queens while already bags of stored chemicals were dotted about the cleared hangar floor. The CFI told me I had an extra week's grace. On Sunday John Fairey rolled his two-seater Spitfire (in which I had my first ride nearly twenty years earlier) above the hangar en route to a display — a happy omen.

Next day, May 5, Vivian Bellamy drove the 240 miles from Cornwall to generously lend his lifetime's flying experience to the venture. (I believe in leaving

The Sport Spitfire's wing being loaded on a flatbed for transport to the airport.

prototypes to the pros.) I took an afternoon off work and we sat drinking tea in the airport cafe watching the wind gusting up to 25 knots. At 7:20 that evening, still in strong winds, Vivian taxied fast along the grass runway a couple of times, said it all seemed OK and opened up for take-off. Twenty-five minutes later he landed after several abortive bouncy landing passes which we attributed to well-known bumps on the runway. He announced a level top speed of 150 mph.

During the ensuing anxious three day wait for weather Viv said he was not at all sure he could get the airplane into the 700 yard airstrip where I had been kindly offered hangar space by John Fairey. However, I checked the tires, inflated by an enthusiastic helper when we first arrived, and rather shamefacedly let out 10 psi; the mystery of the bounces was solved.

On Thursday evening I bribed a forklift driver to move stacked piles of chemicals to make a clearway for the very last airplane to be pushed out of the hangar. It was now or never. After watching two, much easier, practice landings at Thruxton, to my great relief I saw Viv take off for the third time and fly the Spitfire over to the nearby airstrip where he landed without difficulty. At last I could relax briefly.

But although the six year building phase was complete, we now had to get down to the serious business of full flight evaluation to find out just what sort of tiger we had by the tail. Would it really prove to be similar to the immortal Spitfire of which RAF Eagle Squadron pilot Lt. Col. William R. Dunn USAF (Ret.) wrote " . . . highly maneuverable . . . yet gentle and easy to fly with absolutely no bad characteristics"?

Port brake pedal unit. Main fuel supply hose in center.

Sport Spitfire assembled at Thruxton. Dismantled Zlin in the background.

Engine mounting, starboard side, shows pressure cowl aft baffles. Note hot air-bleed box for carburetor below rear cylinder.

This view shows flat bottomed wing blended into curved fuselage by fillets.

Flying the Isaac's Spitfire

By Bob Cole

Thruxton airfield in April and May 1975 hangared two of the most interesting light aircraft to be built in England for many years. They were a replica S5 of Schneider Trophy fame and, appropriately, a 6/10th scale Spitfire. I do some part-time flying instruction and I used to see these machines in their final stages of construction when in the hangar.

The Spitfire flew successfully on 5 May in the capable hands of Viv Bellamy, one of the doyens of light aviation in England. Then I was asked by one of the other pilots at Thruxton if I was interested in flight testing the Spitfire. Could any pilot resist such an offer especially one having already been privileged to fly the real thing?

I met John Isaacs and Viv at the 700 yard strip where the aircraft was then housed, the Thruxton hangar having been filled with fertilizer. John had just decreased the tailplane incidence from -2 degrees (up) to 0 degrees. Viv flew the aircraft and pronounced it much improved. He then gave me a few pointers — the stalling speed and recommended a curved approach — before departing to his flying club at Lands End. I had a problem and a responsibility; flying the creation of six years of a man's life should not be undertaken lightly.

The following weekend I returned to the strip armed with a few ideas, a knee pad, test cards and a stop watch. I had also taken the time at work to pick the brains of John Norris, the trials officer with whom I have been working closely on the Hawk. (The Hawk is a single engine jet trainer built by Hawker Siddeley Aviation for the Royal Air Force.) We had worked out a program for stability checks, performance measurement, take-

Flt. Lt. R. D. (Bob) Cole, test pilot of the Sport Spitfire.

(Photo by Rodney C. H. Poynton)

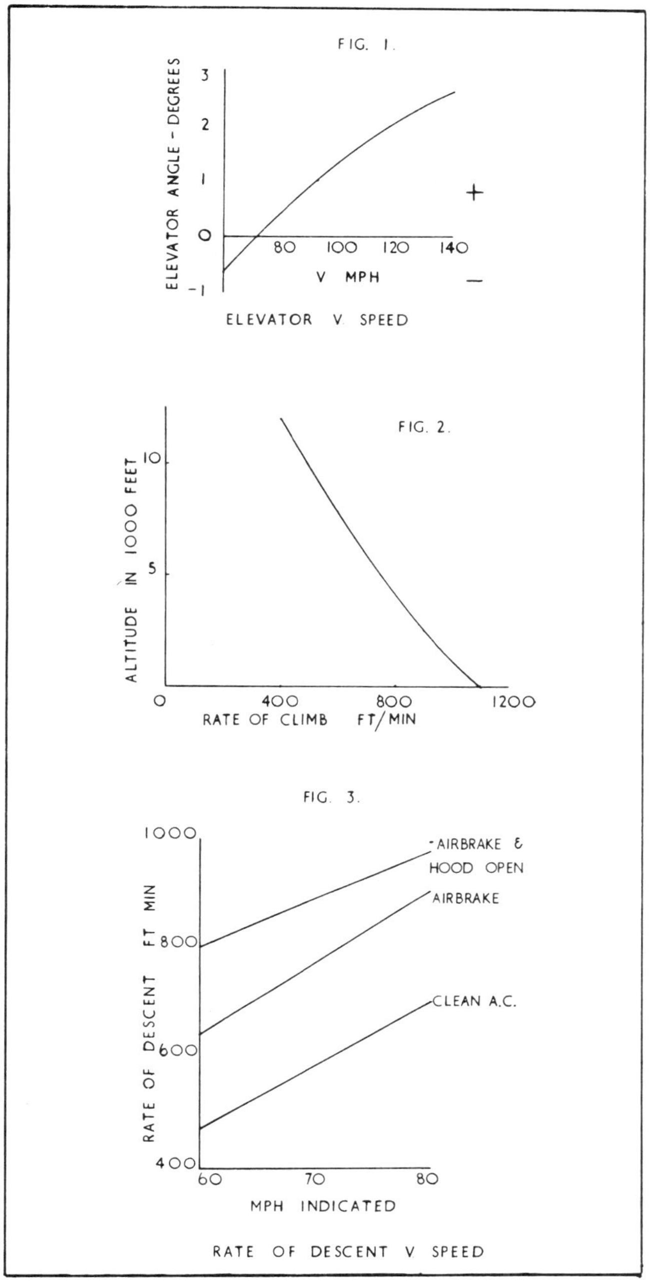

off and landing, stalling, spinning and aerobatics.

On the first day I flew the aircraft 3 times, giving a total flight time of 2 hours and 10 minutes. The first flight was used for familiarity, stability checks and a few stalls. The subsequent flights mainly consisted of performance measurement and stalling with a quick look at the incipient stage of the spin. I could not resist a gentle barrel roll on the way back to the strip at the end of the day.

After this flight testing went on steadily until the Popular Flying Association's International Rally at Sywell on 4th July where John was triumphant in that he won both the Air Squadron Trophy for the best homebuilt and the Tiger Club Cup for the best Original Design.

We now knew what sort of machine we had and what modifications were required. The main problem was the lack of drag when landing; the speed on finals was not easy to control precisely and the float was prolonged. This was solved by fitting a fuselage-mounted barndoor airbrake beneath the cockpit. The airbrake was very effective, with little trim change on selection and a desired approach speed could be flown; the float was much reduced on landing. A further decrease in the tailplane incidence to +1 degree (down) gave a more satisfactory trim speed of about 100 mph with 2000 rpm set.

What else did we find out? Steady sideslips, turns on one control, and stick fixed and free control inputs were made in all 3 axes at 70, 135 and 150 mph. The aircraft stability was normal. The control forces were light but as the controls were well harmonized the aircraft could be flown accurately with little effort. The rate of roll was timed through 360 degrees and averaged out at 90 degrees/second.

Although possessing fighter-like longitudinal stability, careful measurement has shown that although the stick movement with speed is small the elevator movement does show positive stability (Fig. 1). The short period pitching oscillation was well damped as no residual oscillation was seen. As no cockpit trim control was fitted the aircraft was trimmed by fixing cord at the elevator trailing edge to obtain trim conditions for various desired speeds for these tests and to check the long term stability. The phugoid was positively damped and had a 30 second period.

The performance of the aircraft was impressive. Paced take-off and landing distances have averaged out at 200 yards and 150 yards respectively, in zero wind on a grass strip. The stall speeds are 47 mph, airbrake down, and 52-54 mph clean depending on the rate of approach to the stall more than change in fuel weight. A maximum level speed of 150 mph can be obtained at 2700 rpm and the aircraft has been dived to 200 mph with control inputs made to check for flutter.

Partial climbs and descents were made at various speeds to check the rates obtained (Fig. 2 and 3). For airmanship reasons I would recommend an initial climb speed of 90 mph to improve the view. However, above 5000 ft. and up to 12,000 ft. 70 mph gives an impressive rate of climb — still 400 ft./min. at 12,000 ft. Similar reasoning has shown that although 70 mph is the best gliding speed, a forced landing pattern can be flown with much more ease if a higher speed is used; again 90 mph seems a good figure.

Stalling was innocuous, there was no real buffet

I count myself extremely fortunate that chance encounter at the local airfield put me in touch with Flt. Lt. R. D. (Bob) Cole – truly the right man in the right place at the right time.

He joined the Royal Air Force in 1959 and has flown such famous aircraft as the Vampire, Canberra, Hunter, Lightning, Phantom, Jaguar and the Victor bomber. In May 1971, he qualified for the Caterpillar Club when he ejected from a Lightning on fire at night over the Mediterranean. Although currently flying as a test pilot at the British Aircraft and Armament Experimental Establishment, Boscombe Down, and devoting much of his weekend to instructional flying on light airplanes he nevertheless willingly and enthusiastically accepted the additional responsibilities of conducting detailed development testing of an unknown homebuilt aircraft – a task he approached with professional dedication.

ISAACS SPITFIRE

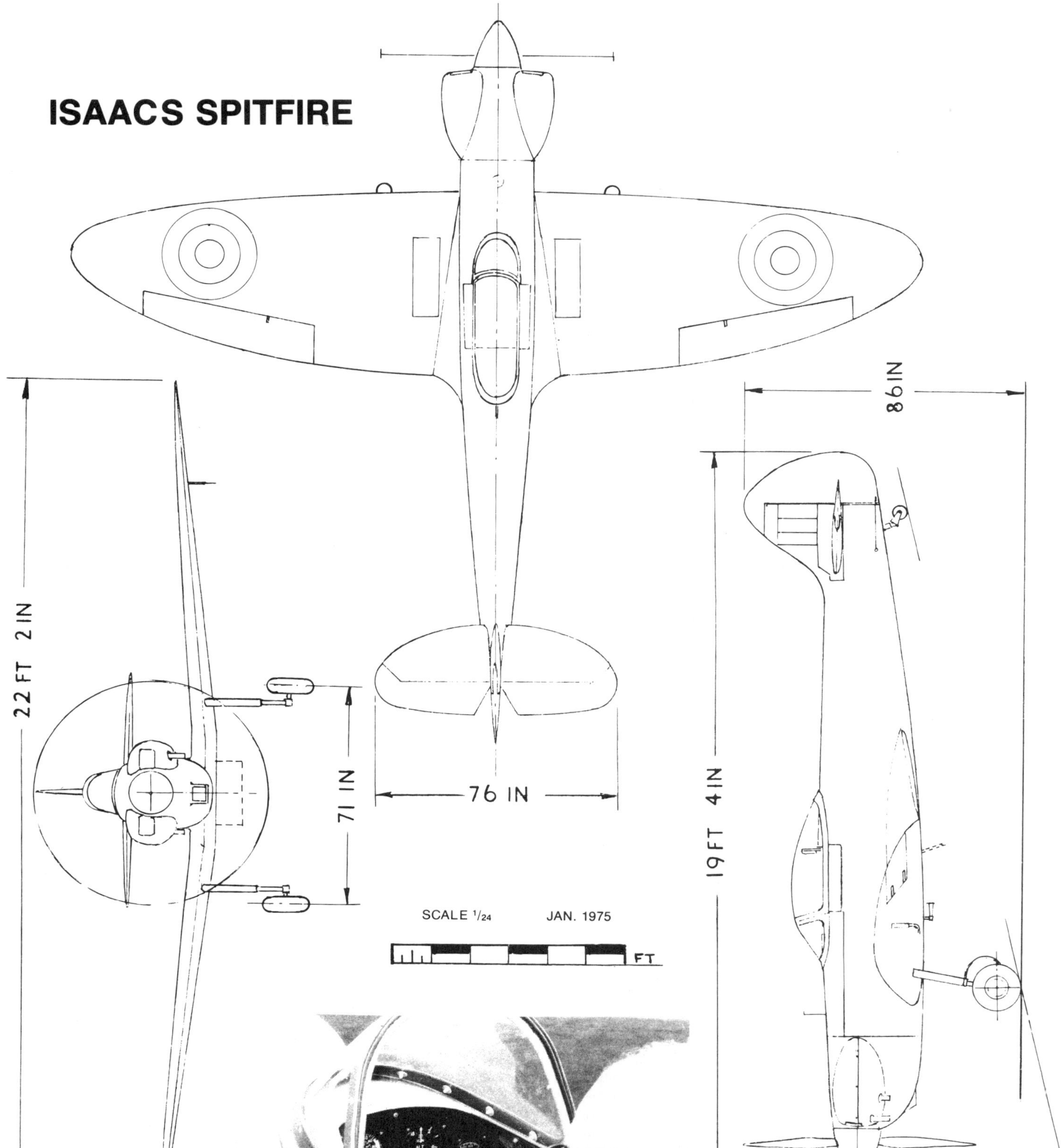

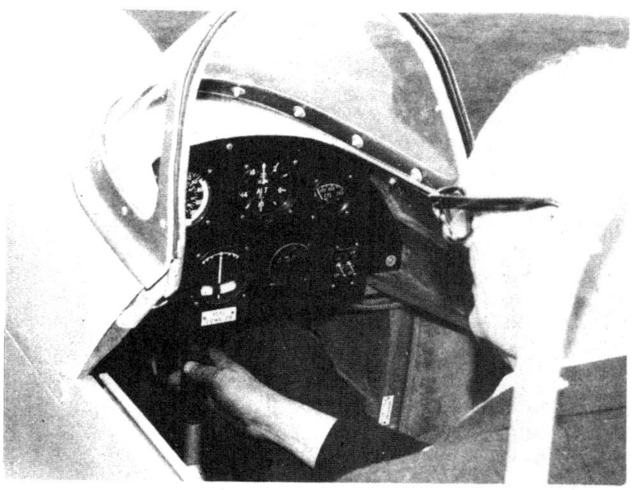

John Isaacs in the cockpit of his Sport Spitfire.

warning but the stall was indicated by the nose gently pitching down accompanied by a slight wing drop. An accelerated stall was indicated by snatching of the ailerons — this can be encountered if you are over enthusiastic doing aerobatics. Spinning on the other hand proved to be interesting. The spin was oscillatory both in roll and yaw, similar to a swept wing jet aircraft. However, a steady spin could be achieved with full outspin aileron. After 2 turns the spin had settled in character with the nose about 50 degrees below the horizon, a full 8 turn spin taking an average of 25 seconds to complete and using up 3000 ft. of sky. Recovery can be made either by centralizing the controls or using opposite rudder; in both cases the recovery will be complete within a turn.

Aerobatics are very easy to execute in the aircraft. All the conventional positive "g" maneuvers have been flown; negative "g" covers the bottom of the fuselage with oil and really does not do the motor any good. It is an absolute delight to be able to perform a continuous sequence of simple aerobatics without height loss or requiring more than 1000 ft of vertical sky. If I was asked to single out one phase of flight I enjoy most in this machine I am sure that I would say aerobatics although, like any tailwheel aircraft a good landing in it is a moment to be savored.

Now imagine yourself strapped securely into the snug cockpit, checks complete and the chocks firmly in position. John calls, "Contact", swings the propeller and you are away. The instruments on the neat panel come to life and the aircraft vibrates gently. Engine checks, pre-take-off checks and then a good look around before lining up on the strip. Ease the power up, watch the swing to the left and then at 30 mph raise the tail. Now we are moving, 60 mph and 200 yards of ground roll ease into the air.

We will hold it low and allow the speed to build — 130 mph, raise the nose, check, then full aileron. Blue sky, plowed fields, then blue sky again — keep the roll going into a gentle left turn to climb to overhead. Less than 2 minutes from rolling we are overhead into wind at 2000 ft. Throttle back, 60 mph snap into a spin — full rudder, stick hard back with full opposite aileron to stabilize the spin. Concentrate, 3½ turns, recover — check 800 ft. Hold the dive for 140 mph before easing up into a loop. You will need a touch of rudder as we come to the top to hold the line. Now relax to make it a better shape, check and roll out as the runway comes into view. Speed and height good so up into the second half of the figure of eight. Out of the second loop so hold the dive till we see 120 mph then quickly up into the vertical stall turn to reverse the direction. Now let us smooth it down, a gentle slow roll followed by a barrel roll. Hold the last part until you see 130 mph before easing into a lazy wing over to reposition. This is also a nice maneuver to show the profile and hide the fixed gear. Full power, hold the dive and wait for 150-160 mph for a final loop; as we come up to the top flick the aircraft using aileron to coax it around. Then coming into the dive again throttle back and check position. A big barrel roll fudging the final heading so we can sweep down onto the runway threshold. The speed is building nicely, 140 mph over the end. So nose up, full aileron and hold the roll to position onto the downwind leg.

It is all plain sailing now. Hood open, carb air on and airbrake out. Peg the speed at 70 mph with a trickle of power if required. A nice tight curving approach onto the runway heading. Roll the wings level and flare. The runway ahead disappears — you might almost be flying the real thing. Feel it gently onto the ground and it's all over. Taxi back to the hangar weaving the nose and unwind mentally.

This sort of flying is like a drug. Words are inadequate to describe the feelings that such an aircraft evokes; admiration for the man who conceived, designed and built it.

(Photo Courtesy John Isaacs)

The Sport Spitfire in its element.

THE ISAACS SPORT SPITFIRE
Specifications

Engine	100 hp Cont. 0-200
Propeller	Ratier (French) Metal
Span	22 ft. 1½ in.
Length	19 ft. 4 in.
Height (Tail Up)	7 ft. 2 in.
Height (Tail Down)	5 ft. 3 in.
Wheel Track	5 ft. 11 in.
Tailplane Span	6 ft. 3.6 in.
Aspect Ratio	5.65
Gross Wing Area	87 sq. ft.
Wing Loading	12.64 lb./sq. ft.
Wing Section	NACA 2200 Series
Wing Thickness	13.2% at root, 6% at tip
Incidence	Root - 2°
	Tip - −½°
Dihedral	6°
Chord at Center Line	5 ft.
Fuselage Airbrake Area	1.16 sq. ft.
Gross Weight	1100 pounds
Empty Weight	805 pounds
Max. Power	100 hp
Power Loading	11 lb./hp
Fuel Capacity	10 Imp. Gallons (12 U.S. Gals.)
Achieved Performance:	
Max. Level Speed	150 mph at 2700 rpm
Cruise Speed	134 mph at 2500 rpm
	123 mph at 2300 rpm
	108 mph at 2100 rpm
	100 mph at 2000 rpm
Max. Dive Speed	200 mph
Rate of Climb	Sea Level - 1100 fpm
	12,000 ft. - 400 fpm
Stall Speed	(Clean) 52-54 mph
	(With Airbrake) 47 mph

(Note: The above data obtained from tests of prototype, G-BBJI. Source: John Isaacs)

How to Cut the Clutter in Your Workshop

by Antoni (Tony) Bingelis

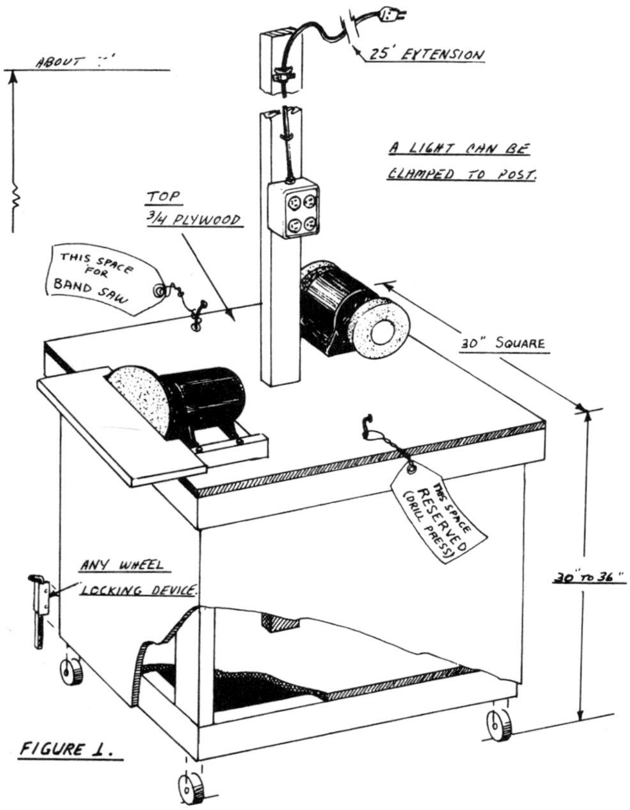

FIGURE 1.

AROUND EAA WORKSHOPS, you find one of two very basic conditions. One, the workshop is crowded and cluttered with the makings of an airplane underway. Two, the shop is empty, pathetically so . . . a project is about to be undertaken but the endless wait for plans or for the ordered materials is being excruciatingly experienced.

Of course, there may be other conditions but I'll wager that the two mentioned fit many more of us. .

No matter what your workshop status may be, you probably can improve the conditions remarkably with a little work, a little effort. If you are already deep in a project, it will be especially hard for you to, even begrudgingly, give up any work sessions to improve things around the shop. It could be worth it though.

Think about the problem for a minute. To build an airplane you need at least enough space to work on the fuselage and, then later, on the wing(s). If space is very limited, it may mean that after you build the fuselage, it will have to be removed to some safe place for protective storage while you attack the wing construction. If you have the work space, naturally, you can afford the luxury of keeping the whole shebang together.

In the building of the aircraft structures, you will be using a variety of materials and some tools and equipment. All of these things occupy space . . . valuable space. They are always in the way, or on top of whatever you need to get, or use, or find. Ever notice that?

The materials, like tubing and longeron and spar stock, are long pieces and have a way of being around, and in the way, for a long time. Two to three years, for example. Then there are the 4' by 8' sheets of plywood or aluminum. Boy! Do they take up space. Last, engine parts, salvaged tubing, cables, pulleys, etc., etc., etc.

In the tool and equipment department you have, if you are fortunate or affluent, a band saw, drill press, bench grinder, sanding disc, bench saw, and a welding rig. But, maybe you may not consider yourself so fortunate after all when you have to provide the space for all of that stuff . . . and your project, too, in one very crowded workshop. Some of it, like the bench saw, may not be used for months on end. Well, that is the prologue. Now for the main event or . . . "How to Cut the Clutter."

THE 4 WAY PORTABLE WORK ISLAND?

There is nothing messier than a bench grinder mounted on a bench. This is true especially where that bench has to serve also as a work bench. Not only does the grinder take up too much valuable space, but in addition, it always has a dirty area around it. The grinder always gets in the way and reduces the usefulness of the bench. If the bench grinder is mounted on a pedestal that is an improvement as it does get the grinder off the bench. One thing about that though, you still have to give up about 3 square feet of valuable floor space for that one piece of equipment.

Everything said of the bench grinder goes double for a sanding disc. It will literally create clouds of wood dust around the bench as the disc rubs away hungrily against expensive spruce. There you have it, two good reasons for the four-way equipment bench. (See Figure I)

Although the bench can be built to any dimension, a 30 inch square configuration should be just about right for most shops. Perhaps, its width should exactly match some existing bench so that you can push it up to that bench and against the wall out of the way.

It is an extremely versatile piece of furniture. You can mount the bench grinder and sanding disc opposite from each other and leave the two opposite sides empty for work space. An alternate and very efficient arrangement would be to mount your band saw and drill press opposite each other in the empty space. If you don't have all of the equipment mentioned, you could still mount any two to four pieces of equipment you do have in your shop and still save space. Some suggestions would include, in addition to the bench grinder, a disc sander, drill press and band saw, a jig saw, 18" sheet metal bending brake, medium-sized vise, small hydraulic press and similar equipment.

The interesting feature is that all of these awkward pieces of machinery and tools can be grouped in just a few square feet of floor space and free your benches of these space robbing gadgets.

The four-square bench can be as lavish as you wish just as long as you make it solid and free from wobble. A central pipe or a two by four is used to mount a convenient 4 plug power outlet. This will give you a power source no matter where you roll the 4 way bench. Shelves and cabinets can be built into each of the four sides to provide storage for drill bits, blades, grinding wheels, and sanding discs, freeing even more space to devote to building purposes.

The central power pole should be at least 7' tall and the 4-outlet electrical box wired with about 25' of extension cord attached so that it can be plugged into the nearest wall plug and still reach to where you are working. Several wire hooks suspended from the workshop ceiling will do nicely in holding the power cord up out of your way. As it is most unlikely that you will be using more than one electrical unit at a time, overloading the extension cord is not even a matter to consider. It would be wise, though, to make sure that the wire is of adequate size for normal use . . . No. 14 wire would be fine and even No. 16 wouldn't be too bad unless you had some heavy load equipment mounted.

Before you get swept up with enthusiasm for the idea, remember that not all of your machines and equipment would be at a uniform working level if mounted on the same surface. Trying to install a bench grinder on a bench with the band saw may result in the working level to be too high for one and too low for the other . . . perhaps you can compromise the heights or even mount the grinder on blocks to help get things at a compatible level. The four-square bench can even have its surface area built on two levels if necessary.

If the 30" x 30" size is a shade small for your purposes, increase the dimensions to suit your equipment. You could make scale templates of your equipment on scrap paper and try arranging the various tools to obtain the most efficient layout. There is no law says you must get four gadgets mounted on this bench. Even two would solve a lot of space problems for you and provide you with a good work island. In mounting something like a grinder, sander or band saw, be sure that no one tool will get in the way of long pieces being worked on one of the other four-square bench machines.

Of course, the real big asset to this rig is the ease with which you can move it about your shop. You can always have the equipment right where you are working. The time you save by not having to walk to each of the machines could reduce the time needed to build your plane by weeks, wouldn't you say?

THE TUBE AND METAL RACK

One of the best space savers you can rig up for a homebuilder's workshop is a rack for tubing and other narrow stuff like long spruce or Douglas Fir boards. This is not a particularly unusual practice but it is a remedy that no builder should ignore.

This really shouldn't take too long to do as you do not have to build a long shelf . . . you don't need long expensive boards. You will need perhaps one or two of these

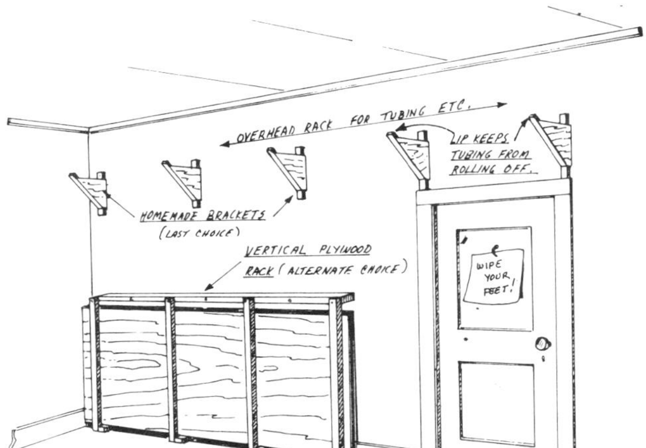

FIGURE 2.

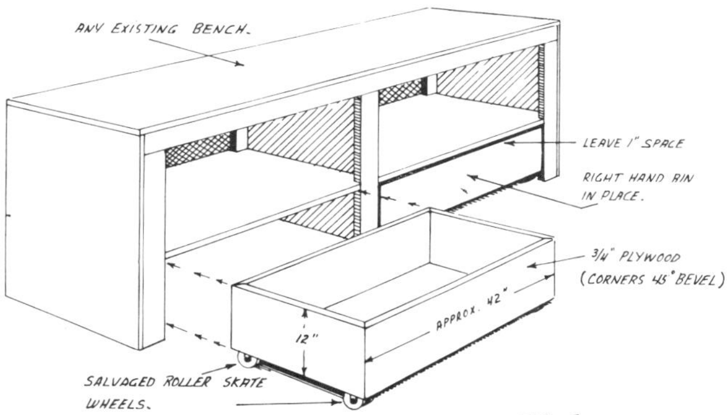

FIG. 3

racks along the garage or workshop wall up near the ceiling where they will be out of the way and not use up valuable wall space at the level where you need it most. A high one above any windows would be a good location.

If you can afford to obtain a bunch of wall brackets and some boards to build long shelves (at least 16' long) that would be great. One drawback to board shelves though, is the fact that you cannot see what you have up there very well without climbing the wall — so to speak. Besides, shelves accumulate an unbelievable amount of dust in a very short time. Long material does not need shelves and several simple wood triangular frames spaced about 4' apart work as well. Take the time to string a taut string from one end to the other (properly leveled) to serve as an alignment for the brackets. This suggestion is not so unique as most builders sooner or later hit on the idea working independently of anyone else. It is just a matter of good clear thinking. A characteristic quite noticeable among builders.

PLYWOOD RACK

Plywood is mistreated in more workshops by more builders than any other material. It is very awkward to handle and to store. As a consequence most gents just lean the sheets against the only clear wall space that may be available. After a few months of neglect, the plywood takes a permanent warp and curved set. That is no way to treat something that costs close to $30 per sheet. If you can't lay the plywood down flat on furring strips so that air is free to circulate under the sheet as well as over it, the next best storage arrangement is a rack nailed to the wall. It takes up the least amount of floor space but you do lose some valuable wall space. A rack suspended from the ceiling is good if you have a sufficient number of cross braces to support the plywood. A wall rack should have outside strips that hold the plywood firmly against the wall to prevent sagging and warping. If at all possible when storing plywood or even structural spruce or fir . . . try to permit the air to circulate around the plywood sheet or board. Changes in humidity cause the wood to shrink and swell. For example, if the air becomes very dry, as in a heated shop in the winter time, the top surface of a sheet of plywood or a board will shrink and the material will curl up. If the same air is permitted to flow across the bottom, the shrinking will be uniform and the warping and curling effect will be just about eliminated.

ROLLERSKATE MOUNTED BENCH DRAWERS

Perhaps one of the most versatile and useful of the shop spacemakers is the dolly mounted drawer. (Figure 3). The space beneath the average permanent workbench is nothing better than a dirt trap where all sorts of things are poked away only to be forgotten forevermore. That alone, should be sufficient reason to salvage this valuable space and put it to good use.

What airplane builder doesn't accumulate many peices of salvaged equipment, tubing, scraps of 4130 sheet, and an unbelievable amount of stuff that may not ever go into the airplane being built. In the meanwhile, where do you store it? Most of the material is **too long** to go into small narrow drawers. It is too valuable to leave laying around and cardboard boxes take up much space and are always in your way, aren't they?

Well, you could be doing yourself a great big favor by taking the time to build a couple of free rolling storage bins or drawers that could be scooted under your bench and pulled out when needed. Mine are mounted on dual roller-skate wheels. It takes two pairs of "cheapie" roller-skates to mobilize both bins. The wheels do not swivel and that is the secret of the success of the bins. They roll in straight and easy once aligned. You can pull them out as far as you want or for that matter roll them clean across the shop if you want. It makes a very neat and nice looking installation. If the drawers are built with about 3/4" space remaining above them when in place, you will find that you can slip material into these long bins without pulling them out. That space also provides the handle for pulling them out. No hardware handles needed.

Being a fair individual, I have foresaken the use of one of the two bins and have it reserved strictly for household junk and hardware. You know, pieces of lamps, light sockets, paint trays and rollers, antenna wire, light switches, brackets, screw eyes, cabinet hardware and anything else that would mess up my workshop for a good . . . but non aviation purpose.

A parting observation. Lawn mowers, edgers, shovels, rakes and other alien equipment have no place in a guy's workshop. One of those small garden sheds or buildings is a wonderful solution, especially if it is far removed from the shop (garage) area where you don't even have to look at the stuff. While I'm at it, I might as well go for broke and emphasize that the workshop is no place to keep the family car(s).

Note: You ladies who sneak a quick reading of the magazine before the old man gets to see it need not write to correct my thinking . . . I have already been counseled (Sigh!)

One of the few present day aircraft to incorporate geodetic construction is the Fisher Flying Products line of kit built wood aircraft. These views show the diamond mesh of thin wood strips on the Fisher wings and fuselage that add amazing strength with minimum weight. Geodetic aircraft have been built by homebuilders in the past, the most noteable being the 4-place Thalman T-4, and Wm. Player's single place Sport parasol which is now in the EAA museum collection. The British Wellington bomber of World War II was of geodetic construction.

Wood - A Forgotten Medium?

by Ben Owen
EAA Technical
Counselor Director

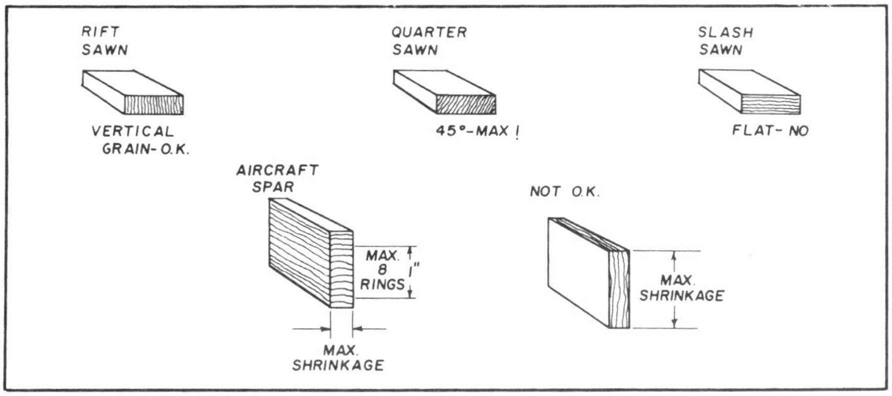

Aircraft wood is best cut with sharp tools, and sandpaper is not necessarily the proper way to finish a surface to be glued. Epoxy works best on reasonably rough, unsanded finishes, and a sharp knife or the sharp edge of a piece of glass makes a better finishing tool than some sandpapers. EAA member John Zark of Streamwood, IL says he spoke to a technician at Chem Tech, Inc. (T-88 Epoxy) since he was using their adhesive, a garnet sandpaper and an aluminum oxide sanding disc on his wood. The technician told him that those are fine as long as the dust is removed from the wood by wiping with the edge of a piece of glass. He also said to clean up any wood that is very rough and hairy (cut with a dull blade) as the epoxy tends to stick to the hairs and does not give a good bond. At least not as good as it could be. The technician also advised against using regular sandpaper and wet or dry paper. It seems wet or dry leaves a residue, which can cause poor bonding.

Aircraft wood is very expensive and should be stored and preserved properly. It should not be set on cement floors or stuffed into the roof rafters without proper flat support. Wood should be supported at least every four feet to keep it from warping, and the ends should be shellacked, varnished or coated with paint to keep it from checking. Fungicides should be applied when storing in a damp climate, and before varnishing. Polyurethane varnish is an excellent finish for wood, inside and out. Wings covered with plywood panels should have the panels carefully placed on the wings and a pencil outline drawn of ribs and spars where the glue goes. These lines should be taped over, and the entire inside surface should be varnished, then the tape removed and glue applied where the tape was before gluing it down. Some thought should be given to preserving the interior of wood structures on aircraft. For instance, since the wing has dihedral, and most of these aircraft are taildraggers, a place where water can collect is against the outside edge of the rib at the front face of the spars, front and back. This is the point at which holes should be drilled and seaplane grommets inserted to help with drainage. It should not be possible for water that works its way inside structures to sit in pools against aircraft wood. Aircraft wood is usually light and strong, but doesn't have much resistance to rot. By the way, there is no such thing as "dry rot", a term used of old as the water had dried out by the time the rot was noticed. Water is required to be present against the wood to give the rot fungus a place to gain a foothold.,

You wouldn't expect in metal aircaft construction to be able to leave a 1/8" gap between materials, nor would you expect to be able to do this with composite structures. Aircraft woodworking does require craftsmanship, and even though modern epoxies can fill gaps, you can see from the prior discussion of the older glues, properly fitted joints are required. It is typical in aircraft construction to discard a lot of parts that simply aren't of high enough quality. That is where the education comes in . . . the recreation comes later! Another part of education is in learning that glues and, particularly, fungicides should be handled with care. They can be dangerous to your health, so please read the labels first.

It might also interest you to know that the military specifications for wood and plywood are not enforced by the government. We have to trust the manufacturer who puts the stamp on aircraft wood indicating it is to Mil. Spec. and that it is indeed of that quality. When the government was checking and approving wood, the specification called for checking at least one in every 20 pieces of wood, and it was still necessary for the user to know enough about wood to be able to select the quality needed.

I recently had the occasion to see some rough cut aircraft spruce that had been improperly stored and supported on about 10 foot centers, it had warped, sagged and its ends had checked because the wood was not varnished or painted on the ends. About $400 worth of aircraft spruce was no longer suitable for aircraft use because of improper storage and care. Some of the spruce lumber we see comes from trees that are hundreds of years old. The supply is extremely limited in aircraft spruce. Quite frankly, it deserves all the respect and care we can give it.

Aircraft plywood is another matter. Quality has to be carefully selected. "Candling" with a strong light underneath thin plywood can help you to detect voids. A major manufacturer of wooden rowing shells (rowboats) fabricates his own plywood from veneer and slow-drying epoxy. One of the beauties of epoxy is that it almost totally eliminates the need for pressure in bonding. Pressure is an absolute requirement for the old glues in bonding. An excellent free two-page article on how to determine the proper pressure for gluing aircraft wood is available from EAA's Information Services office (send stamped self-addressed envelope) if you decide on using other than epoxy.

Getting Started

by Tony Bingelis

EACH NEW PROJECT affects me the same way. I get this feeling that I need to get started immediately . . . to hurry up and start making everything at once even though the plans have not yet arrived. Fantasies run rampart through my mind and I find myself resolving to take my time with this one. That I am really going to build it right . . . build it light . . . and build it pretty. Ah, but that is what I always say and I'm not too sure that these images aren't anything more than builder mirages. Still, a guy should set a realistic standard of quality for his construction so I decided that I would at least strive to maintain a plus or minus 1/32 inch tolerance for my woodwork. That took care of the fantasy bit but I still had to get started.

Does it really matter where you start? After all, everything has to be built and completed before the airplane flies, right?

No Material — No Construction

Since you cannot start building anything without materials, the acquisition of materials must be the starting point of any project. Unfortunately, at the present time the Falco plans do not include a Bill of Materials needed. No matter, nobody initially orders enough of anything anyway when building an airplane.

For me, the materials and supply situation is the same today as it was years ago, a half a world away in Japan, when I started my first Emeraude project. I still have to order almost everything from California, Illinois and from cities other than my own. With the exception of a few hardware items and a limited stock of replacement parts for Cessna aircraft, dear ol' Austin, Texas USA is probably no different than your own locale. If you live in a non-metropolitan area or town far removed from aircraft and aerospace factories, you will know what I mean.

Spruce and birch aircraft plywood are two basic ingredients required for the construction of the Falco. A third, of course, is glue. And that is just about all it takes to construct and assemble all of the structural components for this beauty.

In what I now realize was a shrewd move, I undertook to buy all the aircraft wood I could find (locally) from other builders who had abandoned projects or who had surplus odds and ends remaining after finishing their own projects.

I found the sitka spruce needs easier to fill than my birch plywood needs. More spruce strips and scraps were located among EAA members than birch plywood sheets. But, since the design requires a number of blocks, even small pieces are very useful. Corner blocks and other filler blocks use up a lot of valuable lumber. It would be a shame to have to buy a large, beautiful spruce board only to cut it into small pieces just to make blocks. But there is no avoiding that unless you can get enough offcuts and spruce scraps from other builders. Believe me, in starting any wood project every scrap piece is worth something to you.

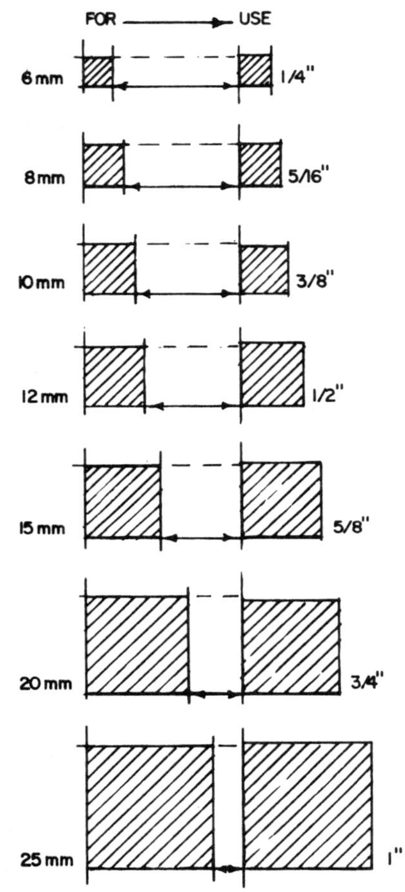

FIGURE 1.

FRACTIONAL SUBSTITUTIONS FOR MILLIMETER SIZES - (WOOD SELECTION)

Anyway, this gave me an instant inventory and at considerable savings. Most of the fellows I dealt with were happy to be reimbursed for just about what they had spent for the wood. With a few recent catalogs in hand you should be able to recognize a bargain when you come across one. Not only are you more apt to buy at reasonable prices from fellow builders but there is also the advantage of immediate availability and the savings in shipping costs. Yes sir, it is a good idea to buy what you can from local EAAers and builders. Better for you and better for them.

What Kind of Glue Are You Using . . .?

In my eagerness to get started, local purchases were made for enough Resorcinol and Plastic Resin glue for my immediate needs (small quantity . . . high price). I also sent off for some Aerolite and epoxy adhesives.

I generally use several different glues as no single adhesive is best for all the different gluing jobs encountered in building an airplane.

I use a lot of Chem-Tech's T-88 epoxy because it is so easy to use a 50/50 mix squeezed from plastic bottles . . . in just the amount you need. No pharmaceutical or laboratory measurements for me, thank you. The 50/50 mixing ratio for the T-88 is not critical. Of course, it like any other adhesive has its drawbacks. For example, this epoxy (and most others) can cause skin allergies in some people if protective measures (skin cream/gloves) aren't used. It can also produce a weak joint if excessive clamping pressure is applied. Furthermore, I hear anew of other shortcomings of the adhesive.

This new round of epoxy revelations making the scene is, as usual, disturbing, confusing, frightening or at the very least, disappointing. Be it as it may, I am again committed. I have already built my Falco spars using T-88 epoxy glue . . . about two gallons of it to be exact (and that stuff is not cheap).

Just so you won't write and ask me what I think of using epoxy glues in homebuilts, I'll expand now on what I have already reported.

A few characteristics of epoxies are well known and each builder should be able to decide for himself if its characteristics are suitable for his purpose. If the designer of your airplane recommends a particular glue . . . why not use it?

It is no secret that room temperature curing epoxies do soften at high temperatures. How high? The number 150 degrees F is one most frequently referenced. Now that is hot! Laboratory tests reportedly prove that the epoxy will get soft at that temperature. Those of you already using epoxy probably already know that and have that proven everytime you disc sand a part that has dried surface glue in evidence. Your electric sanding disc quickly becomes gummed with the epoxy softened by the surface heat generated in sanding.

How hot does it get inside an airplane? How hot will the airplane's wood structure get? Isn't that the big question? I don't know, but the VariEze's epoxied structure operating in the Mojave Desert should get quite hot. However, nowhere near as hot as it would if the aircraft were painted in some dark color. Prove this for yourself by placing one hand on a white car roof top on a sunny day and the other hand on the darker color. Big difference, isn't there?

My reasoning is that wood is a poor conductor of heat and I doubt if an aircraft's structure (wood or composite) would ever be heat soaked to a temperature approaching 150 degrees F. At any rate, composite aircraft appear to suffer no bad effects from high temperatures so why should we expect anything different of a wood structure assembled with epoxy?

But there is more to this latest epoxy flap. There is the claim that epoxy (T-88 included) is not waterproof. Here again, just how waterproof is waterproof and under what conditions?

I have run "screw driver and hammer" tests of my own through the years. Recently, after soarking a couple

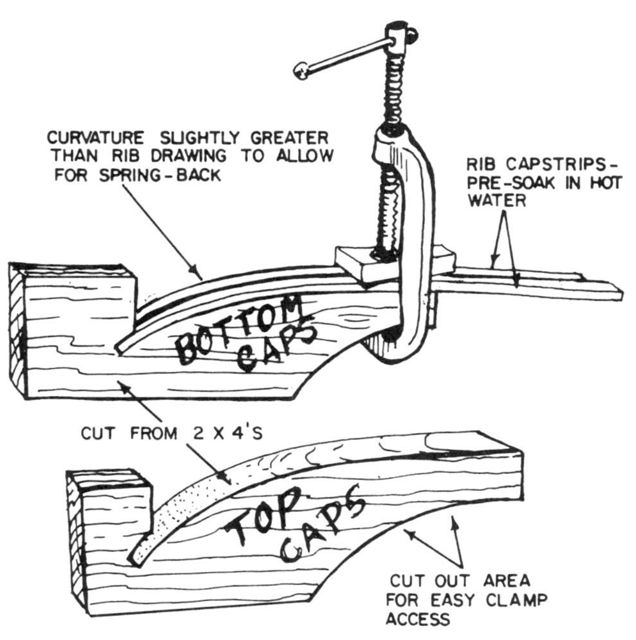

FIGURE 2.

CAPSTRIP BENDING JIG (IMPROVED VERSION)

of Falco laminated bulkhead scraps in water for 50 hours, I clamped them in the vise and broadsided the laminated wood with hefty blows of a short handle five pound sledge. The wood ultimately broke, but about 5 percent of the end portion of one lamination did appear to separate at the glue line. That small failed area might have been due to excessive localized clamping during gluing, or, it might have been a small contaminated area, or maybe the water did affect that portion. However, since 95 percent of the laminated area's integrity cannot be challenged, I dismissed the idea that my epoxied joints are not waterproof. I also recall a couple of plywood and Plexiglass test samples which were glued together with epoxy and set on my gateposts. They remained exposed to Texas weather for almost three years during which time the unprotected mahogany plywood suffered considerable decline but the glued area remained intact to the end. Besides, I think it is customary to protect wood structures from water absorption by varnishing all wood surfaces with a couple of coats of polyurethane or marine varnish. For my part, the waterproof dispute is less a concern than the epoxy's proneness to softening at high temperatures.

But there are other good glues also and if you have any reservations about using epoxies consider using Resorcinol or Aerolite. Both are presumed to be waterproof and unaffected by high temperatures. However, they too have their own quirks. Resorcinol requires good fitting joints, high clamping pressures (about 150 psi) and must **not** be used in room temperatures below 70 degrees F. The other glue, Aerolite, requires application of a colorless liquid catalyst to one of the surfaces being glued. In the course of building the entire airplane, a builder may forget to apply the catalyst at one time or another. The result would be an ineffective joint. However, some ingenious builders add a little cake coloring (I prefer a delicate shade of green) to the catalyst as a tell-tale means of quickly determining whether or not the catalyst was added. Aerolite is very effective for use on ribs and small parts because it sets quickly and requires virtually no clamping.

Capstrip Bending Jig (Improved Type)

The Falco, as is true of most other wood designs, requires that the wing rib capstrips be steamed before bending. Actually, I don't know of any builders who really steam their capstrips to make them pliable enough for bending without breaking. They, instead, soak the capstrips in hot water until they are thoroughly wetted, and then clamp them overnight in a bending jig to dry. The bending jig illustrated in Figure 2 is a big improvement over the old type shown in some old references. You can make a similar bending jig out of a 2 x 4 or a 2 x 6 if it's a highly cambered airfoil you are using. The jig will accommodate two capstrips at one time. If the contour for the bottom capstrips is difference you can make a separate bending jig to accommodate that curvature.

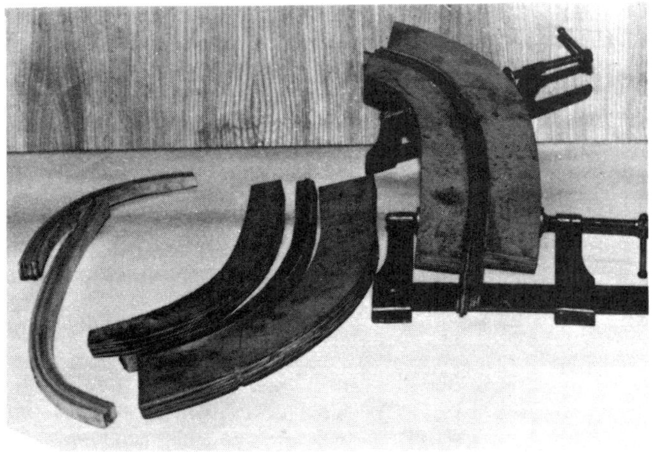

The curvature was so severe for the fin-to-fuselage bulkhead that the individual laminate strips had to be soaked in hot water and clamped to dry. The pre-curved strips were then coated with glue and reclamped to form the separate bulkhead laminations.

What About Working In The Metric System?

Quite a few popular European designs are built utilizing the metric system of measurement and the Falco is no exception. Not only is the metric system used to establish structural shapes but also to dimension materials like wood and plywood.

Working with the metric system is really a pleasure. It is a simple system to use. The only drawback I find is that it is still difficult for me to visualize metric sizes. For instance, I know how big a one inch by one inch block is but I can't quite visualize how big a 30 x 30 millimeter block would be. But, I suppose in time, even that won't bother me. All I really need is a metric scale. If the plans call out a 20 millimeter dimension, I simply measure off 20 on the metric scale . . . neat and no fractions.

Buying wood for the Falco introduces a bit of a problem for builders who want to purchase their wood milled to the exact metric sizes called out in the plans. Unfortunately, most supply sources stock material only in fractional sizes. What is a builder to do if he can't obtain the exact metric size for longerons, capstrips, etc.?

Assembly of the fuselage to fin bulkhead/spar. A simple nail jig was all that was necessary to hold the laminated pieces in proper alignment. Note the use of spring clamps as epoxy requires very little pressure.

If the builder isn't prepared to mill his own stock he can generally substitute equivalent U. S. fractional sizes for most metric sizes shown in the plans. The difference will usually be less than 1/32 inches over or under. As for my own project, I would be very happy, as I mentioned before, to be able to maintain a 1/32 inch tolerance in making and assembling my wood parts. The relationships between metric and fractional sizes that I use are shown in Figure 1.

More of a problem to me was the difficulty in finding some of the plywood sizes required. The Falco's fuselage is covered entirely with 2 mm (about 3/32") birch except for some 2.5 mm (still about 3/32") skin along the fuselage sides. These sizes are readily available as are the 1.5 mm (1/16") plywood skins for the tail and part of the outer wing. It is not until I ran across the need for 1.2 mm (not 1.5 mm mind you) for some bulkhead skins and the 3.5 mm wing spar plywood that I got a bit frustrated. The solution? I laminated a 2 mm sheet with a 1.5 mm sheet to obtain the necessary 3.5 mm thickness. This, however, is not a practical or easy thing to do. Besides it takes much expensive glue to accomplish. Clamping becomes another problem too, to say nothing of the waste of time that ensued.

Nose ribs are being glued to the elevator spar here using rubber bands looped around the leading edge strip and held in place by sticks poked through the rubber band loops. On your left is the workbench with a short 1 x 4 nailed to it so that the elevator can be supported on it and a matching installation at the far end.

Where the plans stipulate 1.2 mm I guess the only thing to do is to go to the next thicker size . . . 1.5 mm (1/16"). Some of my 1.0 mm sheets did mike out to close to the 1.2 size but I wouldn't make a fetish of the problem. Oh yes . . . birch plywood **is** made in 1/32" (1 mm) thick sheets and is 3 ply at that!

I didn't bother to make a materials list for spruce and plywood because I don't think I could have figured out how much I'd need at the start of the project. I, therefore, merely ordered all that I felt I could afford at one time. It turned out that my guess fell slightly short for sitka spruce and way short for plywood. It is surprising how much plywood is required for bulkheads, ribs and skins.

In that the plywood is sold in 4 foot square sheets (some are 50" x 50"), you can rest assured that you will gain a lot of scarfing experience. The curves of the fuselage are much easier to handle with shorter 4 foot sections of plywood so I doubt that it would be worth my while to get any 8 foot sheets for the Falco. The longer sheets could be helpful in skinning the wing but I guess I will stay with the smaller sheets as they are easier to scarf on my bench scarfing rig . . . easier to handle and store, too.

About Profiling Wood Structures

Never attempt to glue plywood skins to a wood structure until its surfaces have been line-sanded to the required contour and trued. Do it carefully and all the joints will be leveled and the squeezee out glue lumps will be removed leaving a perfect surface for skinning.

Truing the surface of a wood structure (line-sanding it) is best accomplished with a long sanding block. One long enough to straddle at least three ribs of the wing structure, or two or three bulkheads when sanding the fuselage. Select a straight board or preferably use one which has been run through a jointer and trued. Glue sandpaper to it on both sides. Use a medium grit on one side and a rough grit on the other. The most durable and fastest cutting sandpaper to use, I found, is that black emery-cloth-like sandpaper used by contractors to sand wood floors. It is stocked in rolls approximately 8 inches wide, so you can purchase whatever length you want. I glue the paper on with Elmer's glue or with Contact cement and then put a weight over it until the adhesive has dried. For my project, I made a 4' long sanding block 4" wide, another about one foot long and a small one approximately an inch wide by ¼" thick and a foot long for use in sanding in hard-to-get-to areas.

. . . And The Laminations!

O boy, does this airplane have laminations! Laminations all over the place! The ideal size strip for most laminating work seems to be about ⅛" (3 mm) thick and in whatever width is necessary for the component being built. All of the Falco bulkheads are laminated eliptical sections. I built mine in two halves and jointed them together with scarf joints. It might strike you as being downright clever to build up the laminations from wider strikes and then rip the glued bulkhead lamination down the middle to get your two identical halves. It might sound good, but I assure you it is not practical. In the first place you would have to rip material into ⅛" strips from a piece of wood more than twice as wide as you need for the completed frame. This is to allow for the saw cut. This also means that if your bulkhead is to be ¾" (20mm) wide, you would have to rip the laminate strips to be at least 1⅝ wide. Also, besides the expense of using wider material, the laminates will be harder to bend and clamp to the jig.

Thoughts On Applying Plywood Skins

When skinning an airplane with highly curved lines like those of the Falco, it is important to try to span at least two fuselage bulkheads with a single piece. This will help maintain a nice curve in the surface skin. On the other hand, if you cover each section with a separate piece of plywood, the effect will be a series of straight lines rather than a nice flowing curve.

This is all the plans show for this portion of the fuselage to tail assembly. Obviously some additional structure must be added to provide support and gluing area for the plywood skins. Note marking for the center lines (leading edges, spar, ribs, etc.). They will come in handy during alignment and finishing.

Clamping trailing edges in place while gluing them to the ribs is a problem because of the difficulty in applying clamping pressure. Here is one way. Small S hooks are bent from welding rod and hooked to rib gussets with rubber bands looped over the trailing edge. Diagonal strips of wood may have to be clamped temporarily to a couple of the ribs to maintain vertical alignment.

The Falco's eliptical frames makes an assembly jig essential. Vertical and horizontal alignment vigil must be maintained with the help of levels and plumb bobs. Each piece of the jig is bolted to permit its extrication once the fuselage is assembled. Note lines on firewall indicating location of internal structure . . . a good practice to follow.

You can force plywood, especially birch, to assume some compound curvature. Simply soak the plywood in hot water until it is thoroughly wetted and then secure it in place, temporarily, on the airplane with nailing strips forcing the plywood's edges to conform to curves beneath. Always start nailing in the center of the frames and work evenly in all directions. After 24 hours the plywood should be dry enough to retain most of its newly acquired compound shape when removed for gluing preparation. It should now glue in place easily and without excessive pressures being necessary. You can either scarf the plywood edges before gluing the panel in place or afterwards. On this project I am gluing in some panels first and then scarfing them in place on the bulkheads using a disc sander in an electric drill. Here again I use sanding discs cut from that black floor sandpaper mentioned earlier. It is a risky way to scarf plywood but it is fast.

Clamps and Clamping in General

by Tony Bingelis

How you can apply clamping pressure to unusual locations calls for unusual improvisation. Here, inner tube loops and sticks solve the problem.

YOU CAN WASTE a lot of valuable building time staring at a problem while you mentally run through alternatives that could solve your most current dilemma. Clamping problems are like that.

Time and again you will be faced with a need to clamp a couple of parts together while you proceed with other work on both of the clamped parts — you know, work like drilling holes, marking, trimming, checking fits and the like. Most of the time the clamping process is routine and you don't even give it a second thought. Other times, nothing seems to work. Your C-clamps are too small or they do not reach in far enough.

Incidentally, have you ever wondered whether you were jigging or clamping? Both terms have synonymical implications. That is, a clamp is any device or gadget used to hold two or more parts together firmly (under pressure usually). A jig, on the other hand, is a device used to maintain the correct mechanical alignment between parts (of work) during assembly. The similarity is great because you could use a C clamp to hold a couple of parts in exact alignment while you weld them. Are you using a jig or a clamp? Who cares! (I just threw that in to get your attention.)

There are no rules for clamping unless it is to use whatever will do the job. On this flap are spring clamps, inner tube loops and nailing strips.

You can't build any kind of airplane without encountering a few clamping problems. Some of these may merely challenge your ingenuity and patience, others can leave you frustrated for days. Understandably, the clamping problems you might encounter could be completely different from those faced by other builders of different aircraft types. Although many of the commonly used clamping devices have a universal use some are designed for very special applications.

There are C-clamps, spring clamps, bar clamps, handscrews, and vise-grip pliers, to name some of the old workshop favorites. These can be used almost anywhere regardless of whether you are building an all-metal, steel tube, wood or a composite airplane. In contrast, cleco fasteners are almost exclusively used in sheet metal work . . . and nails and nail strips more effectively in wood than in metal.

Not to be overlooked as problem solvers are a number of materials that can be employed for clamping work. In this category are rubber bands and inner tube loops, coiled springs, any kind of weight (sand bags, lead and iron scraps), Wood wedges, straps, rope and, of course, screws, nails and bolts.

Also useful as clamping devices, although we seldom think of them as such, are bench vises, woodworking vises and those versatile small drill press vises.

Although I have illustrated (with drawings and photos) a number of interesting clamping applications useful to any builder, in the interest of greater clarity some information regarding the use of the more basic clamping methods is better explained than illustrated.

That pneumatic staple gun allowed this plywood section to be secured in 5 minutes flat. Preparations took 2 hours. Does that tell you something?

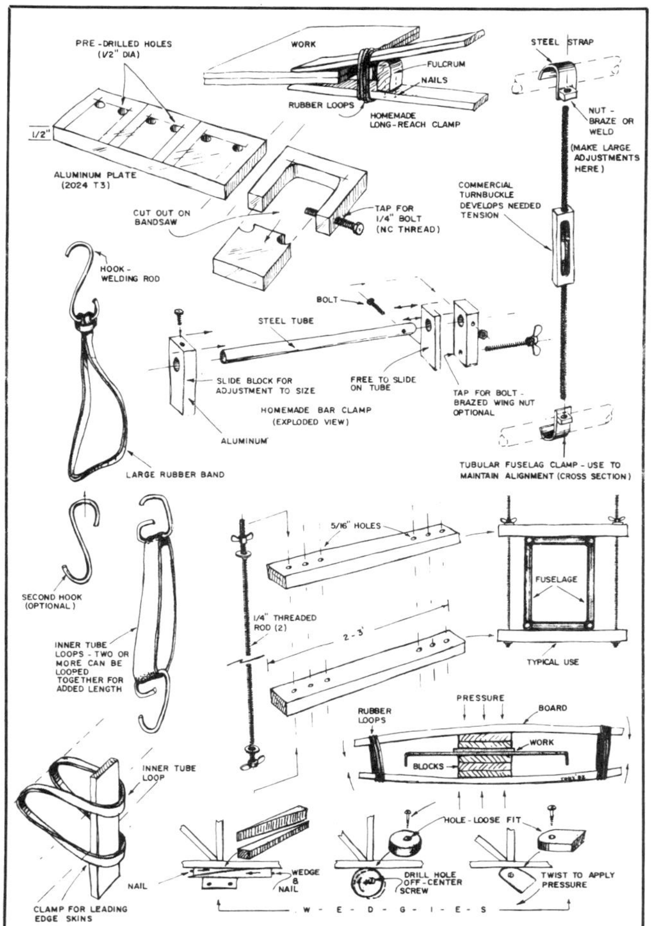

HOMEMADE CLAMPING DEVICES FIGURE I.

Note the back-up blocks to protect the cap strips being bent after soaking in hot water. C-clamps have easy access to such jobs.

Do you see that small Japanese bar clamp and the spring clamp? These excell for the particular needs encountered here.

Clamping Pressures In General

An excessive clamping pressure imposed by any type of clamp is most undesirable. In extreme cases the clamp might be broken . . . particularly one of those "cheapies" we buy just to augment our good clamps. But don't think even for one moment that you can't break your best clamp by injudicious torquing. You can. Excessive tightening of a clamp is more likely to result in surface damage, particularly if you fail to slip a protective block or pad between the clamp and your work. I know you don't always do so but you should — and so should I.

Very obviously the amount of pressure you can impose with a clamp will be, in large measure, related to its size and the type of material on which it is used. Soft materials such as wood can be crushed easily so don't be misled by the resistance you feel (or don't feel) when you are tightening the clamp screw.

Clamping Pressures In Wood Joints

According to that old aircraftsman's "bible", CAM 18, glued joints in soft woods like white pine and spruce require clamping pressures of from 125 to 150 psi and in hardwoods from 150 to 200 psi. These specifications do not necessarily apply today except in cases where similar glues are used. Unfortunately, nobody has ever explained to me how I could determine when I had 150 psi, or whatever, imposed by the clamps.

The epoxies most builders use today require very little pressure. Only enough to bring the two surfaces into contact with the glue line. Actually, if excessive clamping pressure is used, too much of the glue will be squeezed out of the joint leaving very little body . . . and strength. Epoxy depends on an unbroken film or glue line in order for the joint to achieve its maximum strength. This is unlike older glues such as the old time casein, or even the currently produced plastic resin or resorcinol glues. Resorcinol glue, in particular, requires an almost perfectly fitted joint and a relatively high clamping pressure (175 psi) . . . and surface temperatures of the material being glued must also be above 70°F. In the winter time you can bring your shop temperature up to 70° rather quickly but it will take much longer for the project you are working on to come up to temperature.

Whatever glue you use should be applied in sufficient quantity to obtain a slight squeeze-out along the clamped joint. If there is no sign of glue squeezed out along the joint you may have a starved joint or you may have taken too long in spreading the glue and mating both surfaces.

After the application of the initial clamping pressure on a large job, say a wing spar, you should go back after the assembly is completed and retighten the clamps slightly. The reason for this is that some of the excess glue will have been squeezed out and the glue film will have leveled. This tends to reduce the thickness of the glue line and ease the initial clamping pressure that you applied.

If you intend to use nails to impart clamping pressure to plywood skins and leave them in the structure permanently consider these points. Nails will eventually rust unless they are made of brass. Then, too, permanent nailing in external skins (wing or fuselage) will work up in time and the nail heads will protrude. That looks ugly, especially under fabric.

If you intend to use staples and leave them in permanently, use stainless steel staples and don't use them in external surfaces that will be visible.

You can use either nails or staples in conjunction with

Even rubber bands can be useful for imposing light clamping pressures. The leading edge strip is being held in place by rubber bands looped around the leading edge and slipped over short sticks.

"nailing strips". These are strips of wood about ⅛" thick ripped from whatever scrap wood you may have around the shop. It is better that nailing strips be ripped from wood that splits easily.

Nailing strips should be pre-nailed to speed up their application. On the other hand, no such preparation is necessary when you use a stapler or staple gun instead. An electric or pneumatic staple gun is a valuable aid in speeding assembly and one will allow you to do as much by yourself as you could with a helper. Use small diameter staples of sufficient length to penetrate the nailing strip, the plywood skin and the underlying structure. A ⅜" x .030" diameter staple is a very good overall size to use generally.

After the glue has cured, the nailing strips can be removed easily, especially if you used easy-to-split wood. Remove the strips by carefully splintering the wood away from both sides of the imbedded staple. Use a ¼" or ½" wide wood chisel and a small hammer. After the debris of the splintered nailing strips has been cleared away, the nails or staples, whichever you used, will be jutting out enough so that they can be grasped with a pair of diagonal pliers and pulled out. After you have removed a few hundred you will have developed a nice technique with a minimum of blisters.

How many clamps do you need? One more than you have. That bowl of popcorn did nothing to speed up the clamping operation.

Nailing strips are best used in areas where regular clamps cannot reach . . . in attaching large sheets of plywood or plywood skins. However, if there is considerable curvature and the plywood is difficult to hold down, you should use ⅝" nails in the strips. They impart greater pressure than do the staples. A location where nailing strips are necessary is along the leading edge of a wing. Nails will bring the leading edge plywood down with sufficient pressure to obtain a good glue line — staples may not.

A Few Tips For Wood Builders

As builders our biggest problems result from our own impatience and hurrying to get the job over with. That tendency to be hurried causes more complications and delays than if we were to proceed calmly and methodically.

Yes, even string will do a fine clamping job in the right place. After this rudder trailing edge was "strung up" the strings were slipped along the trailing edge to "higher ground" to tighten them.

For example, when you have a big gluing job to be done you should think the process through and prepare for it down to the last detail.

- Find or make all of the clamps and clamping devices you will need, or don't even start the job.
- Open each and every clamp to the approximate size opening needed for its intended location. Locate each clamp near where it will be used.
- Prepare the many little back-up blocks you will use to protect the structure from the clamps.
- Place wax paper any place it might be needed to keep from permanently bonding your work to the bench and everything else. In some places it might be useful to lay a strip of duct tape to stop the dripping flow down the sides of the fuselage. Run it along the bottom of a glued joint.
- To save yourself from a lot of unnecessary work later, wipe away as much of the squeezed out glue and glue runs that you can before the glue cures. It is easier to scrape away with a wood chisel if you allow it to partially set and get thick (not hard).
- Be sure that you can mix the glue, spread it and position the clamps all in the time you will have before the glue starts to set up. If you are unsure, take the time to get somebody to help speed up the process.

Laminating and Bending Wood

by Tony Bingelis

A laminated wood component is a marvel of lightness and strength and it will conform to the required shape better than a similar bend made in a solid wood piece.

Bent aircraft structural components such as frames, bulkheads, longerons, control surface trailing edges, wing tip bows and windshield bows, when made of wood, are almost always laminated because solid wood pieces are difficult to bend.

What Is Laminating?

A laminated wood part is much like a plywood part with this shade of difference. Plywood is a glued up sandwich consisting of three (or more) veneer layers in which the grain of the center layer is oriented approximately 90 degrees to that of the two outside layers. A laminated wood part, on the other hand, is a glued piece consisting of multiple thin layers of wood with the grain of all the adjacent layers running approximately parallel.

The thickness of the individual wood strips to be used in a lamination is governed by the degree of curvature (radius) to which the lamination must be fabricated. Essentially, you should select a single strip thickness that allows it to be easily flexed around the required bend without fracturing. If you find that single strip hard to bend around the required bending form, think how much harder it will be to bend the total number of strips, making up the lamination, at the same time. Obviously then, the fewer the number of lamination layers, the easier the layup will be. On the other hand, such a lamination (two or three layers) will also have a greater degree of "spring back" than one made of many layers. But worth considering too is the fact that the more lamination layers used (due to thinner strips, for example) the greater the amount of glue and wood that will be required to complete the lamination.

Glues

There are many glues that can be used for laminating wood parts. However, without researching specifications and data for various glues it would be better to stay with one of the proven glues used in aircraft construction.

In the category of preferred glues for fabricating laminated bends, we have the urea formaldehyde adhesives (resistant to moisture) and the phenolics or recorcinols considered to be waterproof and known for their strength and longevity. They also perform well in high temperatures.

Use a glue that will allow you a reasonable length of work time before you have to close and clamp the bundle. In this respect, recorcinol is an excellent glue for laminating. Plastic resin isn't bad either, but it is classified as water resistant and not, technically, waterproof. This is, however, nothing to worry about unless you intend to have your airplane double as a submarine, too.

I would not use, or attempt to use, Aerolite glue for laminating work. It is a good two part glue system but it requires extremely fast work because it "grabs" immediately when the two surfaces are joined. When assembling a bent lamination, the strips tend to slide as the clamping and bending proceeds. This sliding action, while of no concern with most other glues, will tend to destroy integrity of the glue line already setting up in the Aerolite.

TABLE: Limiting Radii of Curvature of Thin Laminations
(Material 1/8" thick at 12% Moisture content)

Species				
Standard Name	Botanical Name	Radius (in)	Radius (mm)	Approx. ratio: radius/lamination thickness
Ash, European (home grown)	Fraxinus excelsior	4.8	122	38
Beech, European (home grown)	Fagus sylvatica	4.5	114	36
'Cedar, Port Orford' (imported)	Chamaecyparis lawsoniana	7.4	188	59
Cypress, European (home grown)	Cupressocyparis (Cupressus) leylandii	4.0	102	32
'Douglas Fir' (home grown)	Pseudotsuga menziesii	7.8	198	62
'Douglas Fir' (imported)	Pseudotsuga menziesii	5.9	150	47
Mahogany, African	Khaya grandifoliola	8.5	216	68
Oak, American White	Quercus spp.	5.4	137	43
Spruce, Sitka (home grown)	Picea sitchensis	5.4	137	43

Source: Extracted from Products Research Laboratory Tables

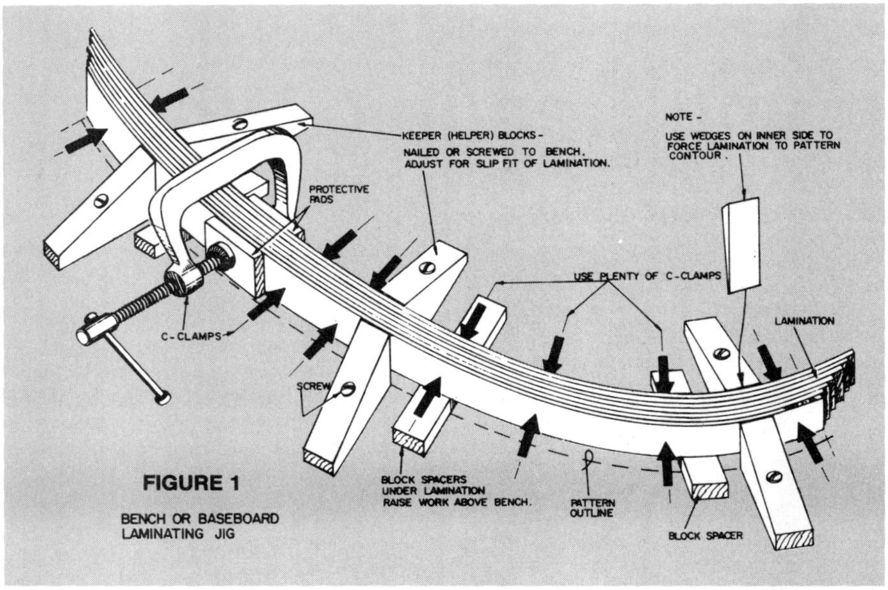

FIGURE 1

BENCH OR BASEBOARD LAMINATING JIG

Laminating Bent Parts

I like to use the blocked-up type of jig (Figure 1) for laminating bulkhead rings, frames, wing tips and similar parts. The small blocks raise your laminating work up off the workbench so that the squeezed out glue has some place to go without sticking the whole works to the workbench. This raised up jig also makes it easier to center the clamps along the length of the lamination.

Here's how to go about it. First, trace the shape for the proposed lamination on the bench full size. The drawn layout is then covered with plastic or wax paper to protect the bench from dripped glue.

Next a few helper blocks are nailed or screwed along the drawn outside perimeter of the lamination pattern. A matching set of facing blocks is then positioned along the inside of the curve. Position these opposite blocks so that the lamination stack will slip in easily. The initial function of these blocks is not to clamp but to merely orient the laminated bundle to conform with the curve while the wedges and clamps are being installed. This means that the helper blocks can be few in number and spaced far enough apart so that you can interspace a sufficient number of C-clamps between them to exert the required uniform pressure on the lamination. Thin wedges are used to force the lamination out to the correct outside pattern contour.

A Dry Run Is Recommended

After checking your jig for accuracy, you are ready to load it with the laminate strips. First, do this without any glue to assure yourself that you can bend the whole batch at one time and still slip it into the bending jig.

As part of the dry run preparations you should position plenty of clamps along the length of the jig where you think you will need them. Don't forget to open up each clamp to the approximate opening required for quick installation.

Also as part of the preparations, include a persuader kit consisting of a hammer and a short 2 by 4 wood block approximately six inches long to use in leveling the lamination during the initial light clamping. Invariably, one or two strips will have an aggravating tendency to rise above the others. These can be brought into line by beating them down (gently, of course) with the hammer backed by the block persuader. Be on the alert for this problem as you gradually apply clamping pressure to the lamination layup.

Clamping Pressures

The amount of clamping pressure that you have to apply will depend primarily on the type of adhesive you use. Recorcinol glues require considerable clamping pressure to achieve a good bond. Epoxies, on the other hand, require a minimum of pressure and a mere intimate contact of the adjacent pieces should be enough. Be especially careful not to apply too much pressure when using thicker epoxies because you may squeeze out most of the adhesive and leave a starved joint that is lacking in strength.

After the initial light clamping is completed (always start in the middle) allow the lamination to set for about 10 minutes and then go back and retighten the clamps lightly as needed. This is necessary because the glue line will take that long to level itself and seep out along the edges.

Making Laminated Bends Using the Insert Method

That is a rather long title but it just about explains the whole process (see Figure 2). This is a good method to use for bending a long, rather straight, wood part that only requires one or both ends to be curved.

The method requires the making of saw cuts in the end to be bent, and so spaced that the wood tongue left between the slots is not apt to break during the bending process. Insert wood strips should be cut to length, coated with glue and inserted into each of the slots. The bend is then completed, as for a normal lamination, in a suitable bending form or jig where it can be clamped until the glue has cured. You can avoid weakening the section to be bent by cutting the slots to varying lengths in order that the insert terminations will be staggered.

Cleaning and Dimensioning the Lamination

After the glue has dried, and before

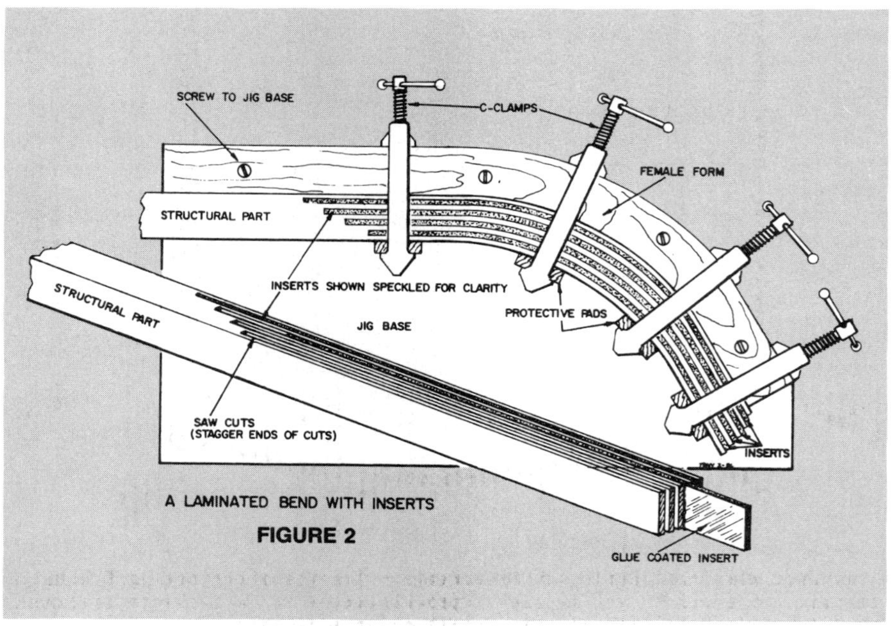

A LAMINATED BEND WITH INSERTS

FIGURE 2

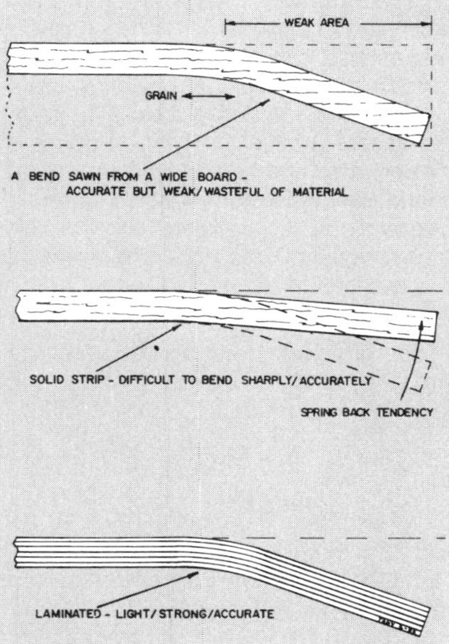

FIGURE 3

BENDS IN STRUCTURAL PARTS

Local EAA types in action critiquing a well built laminated rudder. Picky picky picky.

you remove the completed lamination from its jig, clean away the dried glue from the top surface. Use a disc sander, or coarse file, to level the glue globs. Follow this with a final clean up using a sharp block plane. Plane away only enough material to clean and level the surface.

Inspect the lamination to see that you have uniform glue lines with no open spaces. If a gap or two is found, these can be filled with epoxy provided they are not extensive or involve several laminations. However, gaps should not be a problem if your clamps were centered and you obtained the recommended light glue squeeze-out during clamping. If you find that you do have an unsatisfactory lamination, discard it. Determine where you went wrong and correct your procedure before attempting to make any more laminations.

Finally, remove the lamination from the jig and then clean the bottom surface as you did the top side. Be careful when using a sanding disc as it can be hard to control. Trim your lamination to the correct width with the block plane or with a long sanding block.

A well made lamination is a pretty piece of work and you, like many builders, will hate to see such good work hidden from view in the completed airplane.

An Alternative To Laminating and Bending?

If a wood part is hard to bend why not cut out the required curve from a wider board? It is sometimes possible to avoid the bend problem that way, but it will result in a part that is more expensive

A laminated wheel well ring is easy to fabricate in this type of plywood jig. Drill large holes about one inch in from the edge to provide access for the C-clamps as shown. Suggest you block the plywood up to simplify clamp installation.

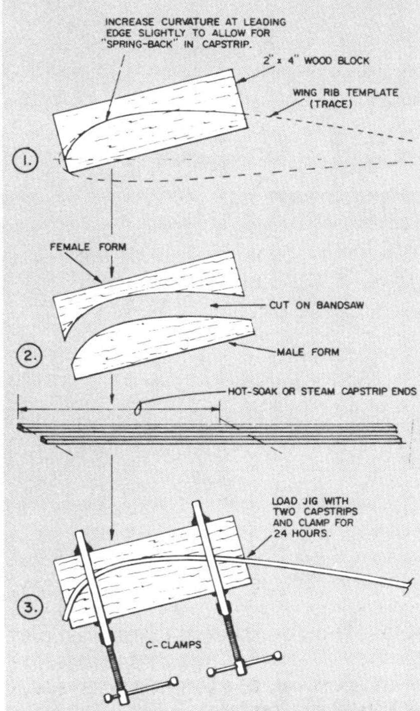

FIGURE 4

BENDING BETWEEN MALE AND FEMALE FORMS

Buzz Glade Photo, Orange Park, Florida
How many laminated parts can you count? If you said 9 you would be close to right. The turtledeck and firewall are laminated components as is the windshield bow in addition to the fuselage bulkheads.

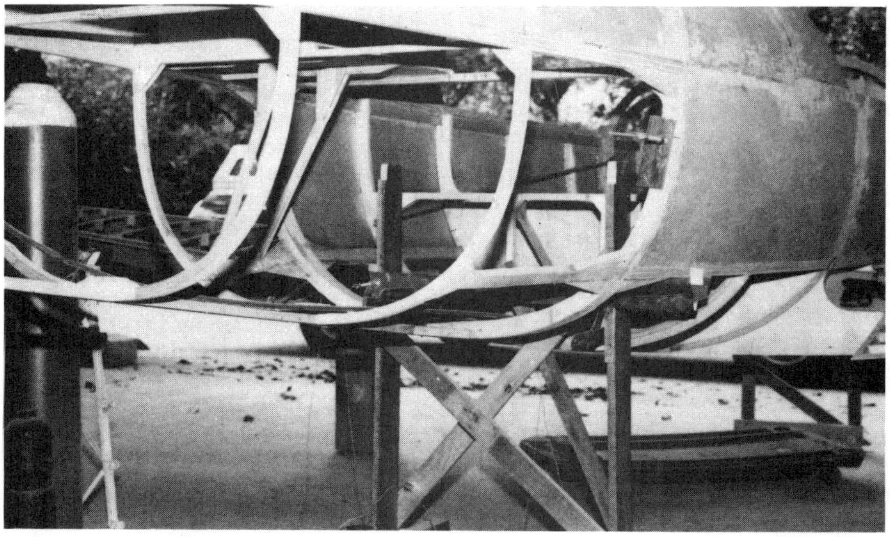

Where would you use laminated stuff? How about in a fuselage such as this. Those bulkheads are all laminated.

Here is the completed wheel well lamination all cleaned up. Ends will be joined through a scraf joint . . . about 4:1 is plenty for this purpose.

and weaker than a similar bent part. The reason, of course, is that the grain in the area of curvature is taking a "short cut" across, and any load imposed across the curve will be where the grain is the shortest and the weakest. Reinforcing this area with plywood gussets will help improve its strength, but what will that do? Add extra weight and bulk! (See Figure 3)

Minimum Bend Radius For Wood

The minimum bend radius will vary from one wood species to another, and also with the moisture content of the wood. The Table accompanying this article will give you a rough idea of what the comparative minimum bend radii are for the selected woods used in aircraft construction.

Studies and tests conducted by the Forest Products Research Laboratory show that out of twenty test pieces bent to the minimum radius indicated, no more than one piece should result in a fracture. The test results are based on the premise that good wood free from defects will be used. So much for the scientific stuff.

Softening Wood For Bending

Solid wood parts like longerons — and especially wing tip bows — are very difficult to bend without fracturing if the bend radius is comparatively small. A dry solid piece of wood will offer considerable resistance to bending and even when successfully installed in a forcedly flexed condition, it will continue to resist that imposed bend for a long time to come. This forced conformance imposes very definite stress loads on the installed member and on the adjacent structure as well. A most undesirable condition.

Any wood can be softened for bending. That is, the plasticity and compressibility of a species can be improved by subjecting it to heat and increasing its moisture content. Wood can best be softened to a semi-plastic state for bending purposes by steaming it. Actually, immersing the wood in boiling water for approximately 15-30 minutes may be considered to be the ultimate softening treatment for the homebuilder. This softening method may be fine for wing rib capstrips but such an accommodation for larger pieces of wood will require a more functional container than a pot. A large diameter steel pipe with one end capped would do. Just add water and heat it and you will have a fine steamer for longer wood parts.

Fuselage longerons requiring a rather severe bend will have to get a localized wood softening treatment. Wrap the longerons where they will be bent with wet towels. Pour on boiling water and allow the towel wrapped

How would you center the wheel well ring during installation? A plywood disc with a centering hole slipped over the landing gear axle work perfectly here.

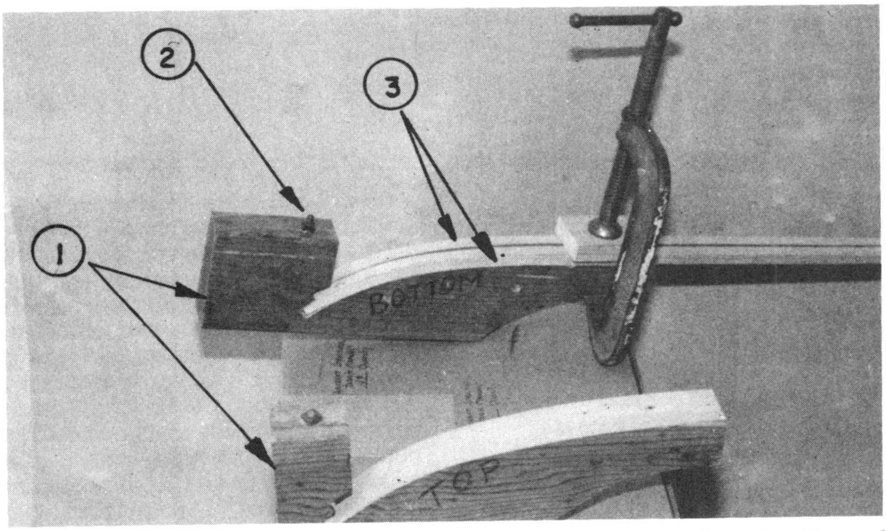

For making simple capstrip bends it is hard to beat this jig for ease of construction and use. 1. A 2" x 4" block cut to the wing rib shape with a bit extra curvature up front. 2. Reinforcement bolt. 3. Two capstrips can be bent at one time.

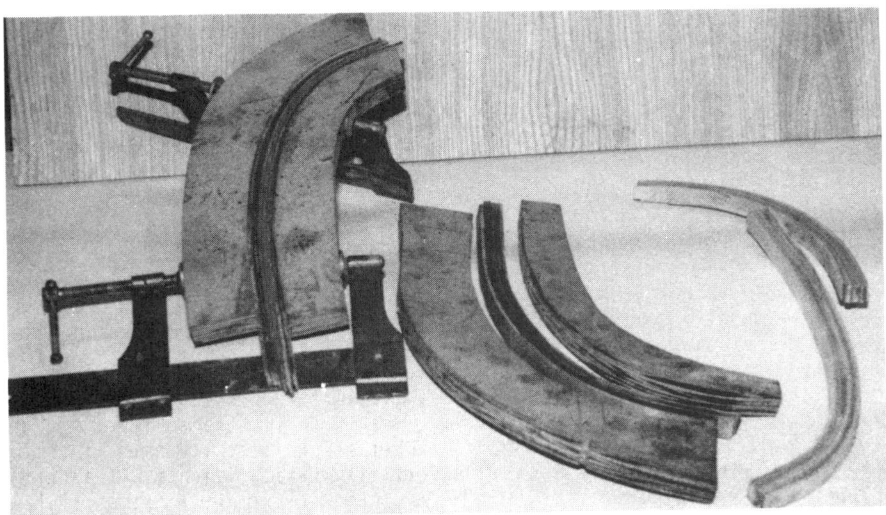

A male/female bending jig need not be elaborate. It is an excellent means for bending or laminating small parts as shown here.

wood to soak overnight. Repeat the hot water treatment just before making the bend. Usually, the bends in longerons are fairly gentle so this treatment should be sufficient in most cases.

Pre-Bending and Spring-Back

Most likely the most common wood bending requirement that a builder will encounter is that of pre-bending of capstrips for wing rib construction.

Pre-bending, as the term implies, is a procedure for imparting a permanent set to a bent wood part before it is used in an assembly. For wood components where the bend radius is severe, this is accomplished by steaming or soaking the wood before loading it in a bending jig where it is allowed to dry.

A part that has been allowed to dry in the jig will, when removed, retain most of its bend . . . most but not all. This relaxation of the bend is due to the natural memory factor in the wood which results in this "spring-back" characteristic.

There is no reliable way to figure how much spring-back to expect. A wood's resistance to bending and its reluctance to maintain a bend once obtained in a solid piece are largely affected by the dimensions of the piece and by its latent moisture content.

Capstrip Bending Forms

Elsewhere one photo shows a very easy-to-make bending form for pre-bending capstrips. Its open design allows the wood to dry much quicker than it would in a male/female pressing form (see Figure 4). The illustrated male/female pressing form is, perhaps, easier to make as it requires only a single curved cut on a bandsaw. Either type of bending form can be made from a 2 by 4 block of wood. This size permits you to load the form with two capstrips at a time.

As for any pre-bending procedure, soak your capstrips thoroughly in hot water or, preferably, immerse the ends in a container of boiling water for about 15 minutes before bending and clamping them in the jig. Allow an overnight to 24 hours drying time before removing the capstrips from the jig.

The pre-bent capstrips will be much easier to install in the wing rib jig and the threat of having one break is eliminated.

THE AIRPLANE ON THE COVER

Ray Downs' Corby Starlet

The superb all wood beauty pictured on the cover represents the first of two Corby Starlets built by Ray Downs, Costa Mesa, CA. N213RD was the first completed and was flown in spring of 1989.

Prior to retiring, Downs operated a hobby shop and was a nationally known radio control model airplane builder and flyer. As these photos show, construction of the Starlet is similar to a scaled-up model airplane. Simply draw out the fuselage sides full scale on the work table, tape down wax paper, and glue up the frames. Only difference actually, is using 3/4" spruce instead of 1/8" square balsa stock — and nailing down form blocks rather than pushing in straight pins to keep longerons and intercostals in place while the glue dries.

Downs built the plane in his one-car garage in the condominium complex where he lives. Note the neatness evident on the walls of his workshop.

The 18.5 ft. wing is built around a single laminated wood spar. Builders have the option of making the wing in one piece, or in two panels. Downs decided on the two panel wing because of the restricted work space in his shop. There is a penalty however, as the two piece wing involves much more effort, is more expensive and adds to the aircraft's weight.

Powered with an HAPI 1835cc VW conversion, Downs' single seater indicates 135 mph at 3000 rpm. Top speed is 160 mph with stall occuring at approximately 48 mph.

Aircraft has two fuel tanks — a 9 gal. main tank forward of the instrument panel and a 6 gal. auxiliary tank behind the seat. Aircraft is equipped with nav lights, three strobes and is fitted with a loran and Mode C transponder. With the big clear sliding hatch, deep upholstered seats, Downs' Starlet is a delight to fly locally or on extended cross country flights. Ray Downs also built a Wittman Tailwind — painted like his Starlet and all models he builds — in International Orange and Bahama Blue.

Plans for the Corby Starlet are available from Vic Boyce, 510 NW 46th Terrace, Plantation, FL 33317.

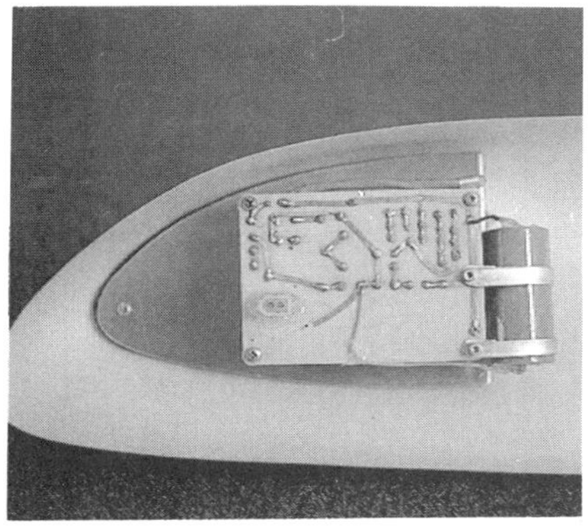

Down's strobe mechanism mounted on the outboard wing ribs. Components were purchased from Radio Shack. After form blocks were made in the shape of the wing and rudder tips, he heated red, green and white plastic in the kitchen stove and worked the plastic over the forms until desired shapes were formed. A neat, inexpensive installation!

Most wood aircraft are just like an oversize model airplane in construction. A bit more complicated to be sure, but basic building techniques are remarkably similar as the tail assembly above, and the partially completed fuselage and wing to the right indicate. The attach fittings mounted on the main spar are additional construction items required when a two-piece wing is made for the Starlet. This involves more weight, is more costly in material and time, but it's a worthwhile trade-off if work space is restricted. Top photo shows VW engine installaiton with baffling in place.